Special Thanks

I am deeply grateful to my parents, whose unwavering support and encouragement, especially during the challenging moments, have shaped me into the person I am today.

They didn't just give me life, they inspired me to live it to the fullest, always urging me to grow, strive, and never settle.

Thank you, Mom and Dad, from the bottom of my heart.

Author: Jonathan Cantin
Copy edited by: Albert Cantin

Mastering CNC and Digital Fabrication II
From Curiosity to Competency in 200+ Questions about CNC Routing, Laser, Plasma, Waterjet, 3D Printing, and Welding

ISBN: 1-896369-54-5 // 978-1-896369-54-9

V1

TABLE OF CONTENTS

CNC routers produce perfectly flat, parallel, and square blocks.

It seems only fitting that when I finally earned my diploma, I made a custom metal frame for it from 1/8 inch wall, 1 inch square tubing. The diploma weighs almost nothing, but the frame that holds it weighs close to ten pounds.

After years of doing everything except metal fabrication, the weight of it is still something I am getting used to.

Foreword
The Curious Maker's Companion
Why This Follow-Up Was Inevitable

You'd think after writing a massive CNC book full of charts, stories, quizzes, and even history, I'd be done. I mean, how much more is there to say about routers, lasers, plasma cutters, waterjets, 3D printers, and welding? Turns out a lot.

After publishing Mastering CNC and Digital Fabrication, I realized that it served more as an introduction than a real conversation, similar to the coursework I did during my two years earning a welding diploma. It touched the surface and left you with more questions than answers. Then business slowed down, I took a welding job, and suddenly I was deep in the metal fabrication world for real.

The goal now is to build on that foundation and make more sense of what came before, because once you understand the why, everything else starts to click.

Remember when you constantly asked "why" as a kid, but somewhere on the road to adulthood, that curiosity faded? In the shop, though, asking "why" is exactly what saves time, materials, and sometimes even your life.

It is how you start connecting the dots between tools and materials. Learning to think like a fabricator. Understanding the reasoning behind the process, not just following steps, keeping you from turning into the CNC world's version of a MIG Monkey.

CNC and general fabrication never stop evolving. Faster machines, smarter software, and new materials bring more potential, but also more ways to screw up. Think of this book as your sarcastic shop buddy who's seen it all and quietly mutters, "Don't do that. Here's why."

Ready to test how much you really know? Take the 100-question Master Quiz **right now** *(p. 303), then again after reading the book. Let's track your learning ROI.*

Jonathan Cantin, The CNC King
(and founder of CNCROi.com)

CNC lasers and routers both deliver high-quality results on lamacoids.

But lasers produce sharper results without hold down issues.

Section 1
The Router Bench
Bits and Passes, Let's Avoid Them!

CNC routing is like giving your workshop a robot assistant, one that doesn't complain, and cuts your designs with laser-sharp precision (well, technically it's a spinning bit, but you get the idea).

Whether you're carving intricate wooden signs, shaping plastic parts, slicing through foam, or even dabbling in soft metals, a CNC router brings industrial-level capability right to your bench.

At its heart, CNC (Computer Numerical Control) routering takes the designs you create on your computer and turns them into perfectly cut parts by controlling the router's movement along the X, Y, and Z axes. Forget shaky hands or mismeasured cuts, the CNC router does exactly what it's told (even if you tell it something wrong).

1.1 Why does industry love CNC routers?

From cabinetry to aerospace to signage, CNC routers deliver precision and repeatability every day. Whether in a huge shop or a home garage, they slot in perfectly.

First, it's fast. What might take hours or days by hand can often be done in minutes. Sometimes it finishes a job so fast, you'd swear it skipped steps.

Second, it's consistent. Whether you're making one part or one hundred, every piece comes out the same. It's like cloning, but with fewer ethics discussions.

Third, it's versatile. Wood, plastics, composites, foams, and even non-ferrous metals all fall under its capable reach.

Basically, it's a workhorse that never asks for overtime pay, and doesn't mind working weekends.

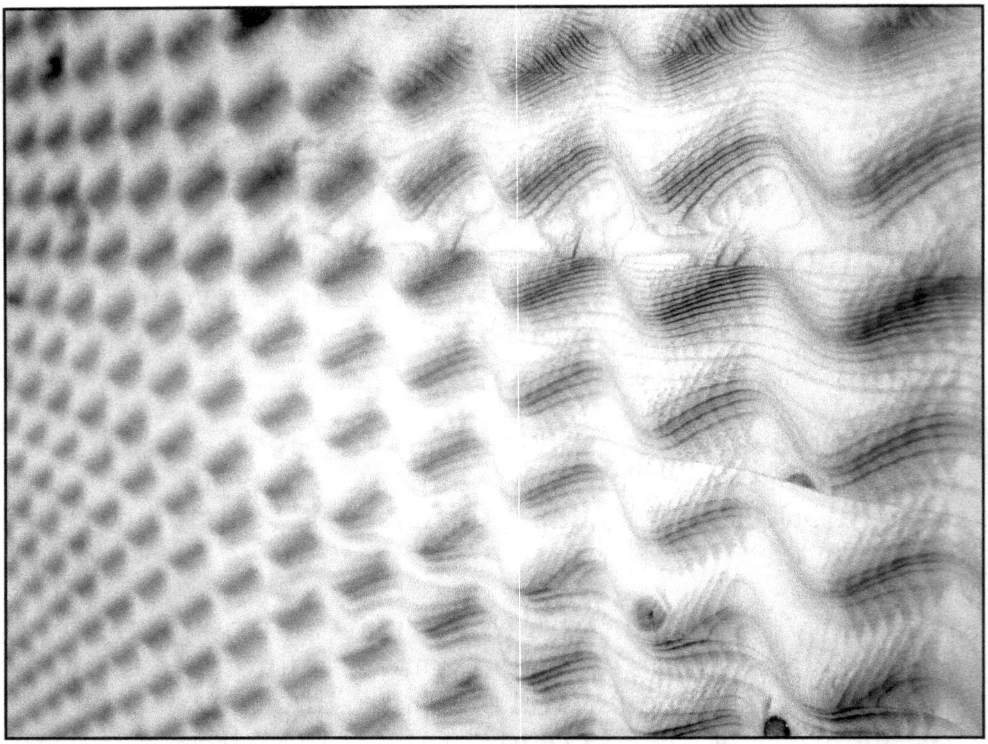

CNC carving delivers unmatched precision.

1.2 CNC: The Digital Backbone

If you're stepping into the world of CNC, a router is hands-down one of the best first machines to get your hands on.

Other CNC machines can feel intimidating, with steeper learning curves and more expensive mistakes waiting to happen.

A router, on the other hand, is approachable, relatively affordable, and small enough to fit in a modest shop or even a large garage. Cheap bits break before the machine does, giving you room to experiment without panic.

You do not need a degree in rocket science or to master 3D modeling just to cut a simple part.

I started with a ShopBot Desktop and watching a flat sheet of material transform into a perfect part felt like pure magic. It is digital wizardry with zero robes required, unless you really want to wear one.

Routers are the perfect gateway into CNC. They introduce you to the holy trinity of toolpaths, feeds and speeds, bit selection, and material hold-down. Master those, and you're basically learning the secret handshake to the bigger, badder CNC machines waiting down the road. Think of it as training on a simulator before flying the real jet, except your simulator is a professional wood chipper.

And let's be honest: there's something deeply satisfying about that gentle whirring as your project takes shape, like your machine is whispering, "I got you, boss." Every pass reminds you that you programmed this thing to dance.

CNC routering is addictive, satisfying, and wildly useful. Whether you're a hobbyist, running a small shop, or just someone who loves the smell of freshly cut MDF, starting with a CNC router puts precision manufacturing right at your fingertips.

Now, let's dive deeper into the lore of CNC routing and see what makes these machines tick.

Glue-ups are an essential skill when working with a CNC router.

1.3 Router Safety Myths & Mistakes

The CNC router: the beautiful beast that turns your digital designs into precise, tangible masterpieces.

But let's be honest, behind the humming motors, whirring spindles, and hypnotic toolpaths lurk a whole bunch of safety myths and operator mistakes that can transform your shop from a place of creative triumph into a site of chaos, and, at best, expensive repairs.

Real hazard in the shop isn't always the machine.

Sometimes it's the overconfident operator who told the router it was cutting 4 inches deep... when the material was barely a quarter inch.

Myth 1: "It's all automated, I can walk away and grab a sandwich."

Spoiler: you cannot. Yes, CNC routers run on G-code and follow precise instructions, but they're not magical elves, they can and will mess up when you least expect it.

Leaving ANY running machine unattended is like leaving a toddler with a Sharpie near a white couch: it only takes a few minutes for disaster.

Tools break, clamps come loose, bits overheat, material shifts, or the spindle decides to plunge through your carefully programmed Z-zero like a mechanical woodpecker on steroids.

Always monitor the job, even if you think it's routine.

Myth 2: "Dust? Eh, I'll sweep up later."

CNC routers are magical, yes, but they're also summoners of fine particulates. That delicate sprinkling of MDF, hardwood, soft metal, or especially Delrin shavings gets everywhere. While it might look harmless, this dust is a respiratory nightmare, and worse, a potential fire hazard.

Without good dust collection, your shop becomes a ticking time bomb of fine particles suspended in the air, just waiting for the right spark to turn the whole place into a grain silo explosion reenactment.

And let's not forget the long-term perks of neglect: gummed-up bearings, clogged motors, shortened machine life... and yes, mushroom lungs, the unofficial badge of shame for those who thought MDF dust wasn't a problem.

Wear a mask. Run the dust collector. Sweep as you go. Future you (and your lungs) will thank you, preferably from a dust-free, non-smoking garage.

Myth 3: "Clamps are optional; the vacuum table's got this."

Yes, vacuum hold-down systems are amazing, but they're not miracle workers. Forgetting to clamp small, lightweight, or irregularly shaped pieces is like assuming your dog will hold still while you balance a cookie on its nose.

Small parts can shift under cutting pressure, and when they do, they often become tiny high-speed projectiles. Use clamps, tabs, or even sacrificial boards to back up your vacuum hold-down and avoid surprise frisbees.

Myth 4: "It's only wood, I don't need ear or eye protection."

Whether you're cutting wood, plastic, or metal, you've got a high-speed spindle, razor-sharp carbide, and chips flying like miniature shrapnel. Yes, God gave you a spare, but what happens after you use that one up?

Even softwoods can launch splinters, and the noise from a CNC router, especially on harder materials, can sneak past safe hearing thresholds faster than you think.

Want to keep your hearing and avoid looking like a pirate with an eye patch? Always wear earplugs and safety glasses, even if the material seems "gentle."

PPE is a must, even if you're working alone in the shop, like I usually am at *CNCROi.com*.

Myth 5: "Emergency stop? Pffft, I'll never need that."

The emergency stop button isn't just a big red decoration on your control panel. It's there for when, *not if*, something goes sideways, and trust me, it will go sideways someday.

Whether it's a tool crash, a software hiccup, or the dreaded "I forgot to clamp the part" moment, having your hand near the E-stop can save your machine, your material, or YOU!

Don't kid yourself into thinking you're a flawless operator. Keep the E-stop clear, accessible, and mentally rehearse using it before every big job. When chaos hits, muscle memory beats panic every time.

Myth 6: "I ran the program once; it's fine forever."

Congratulations on your successful first cut! But don't assume you can run that exact same file next week, on a different material, with a different bit, and expect perfection. Even identical twins aren't identical.

Even the roughest wood can become beautiful.

Material density changes, tool sharpness changes, even room temperature can affect machine performance (seriously, thermal expansion is a thing).

Every job deserves a careful check of the program, the toolpath, and the material setup. Skipping this step is the express route to either "why is the bit buried halfway through my spoilboard?" or, worse, membership in the *Missing Fingers Club*, the initiation ritual is brutal.

Myth 7: "I don't need to resurface the spoilboard, it looks totally fine."

The spoilboard: the unsung hero of every CNC router setup. It's the thing you cut into so you don't damage the machine bed. But over time, it warps, grooves, and turns into a lumpy landscape of unevenness.

Ignoring your spoilboard means you're inviting uneven cuts, inconsistent depths, and headaches galore.

Regularly resurface it with a fly cutter, keep it level, and treat it like the precision surface it is, because the router only knows what Z-zero tells it, not the true high and low spots of a warped surface.

Myth 8: "The more flutes, the better the cut!"

More flutes = smoother cut, right? Wrong. While extra flutes on a tool can provide finer finishes, they also reduce chip clearance, especially in soft materials like wood and plastics.

This can lead to overheating, melted chips, and a gummed-up, burned mess. Choose the right tool for the material, not just the fanciest one. Remember: it's not a beauty contest, it's about moving chips efficiently and safely.

Myth 9: "I can just hand-edit the G-code; no need to re-export."

Look, you're clever. You know how to open a text editor, spot the offending Z-depth line, and change that pesky number from -6 to -4.

But here's the thing: one slip, one wrong axis letter, or one missing semicolon, and you've just programmed your machine to plow sideways into a clamp or pierce the table.

Hand-editing G-code is like defusing a bomb with gardening gloves on: sure, you might get away with it, but you probably shouldn't. Use the CAM software properly and save yourself from a pink slip.

Myth 10: "Maintenance? Bah, if it ain't broke, don't fix it."

Famous last words. CNC routers, like any precision machine, need routine care: greased bearings, clean linear rails, properly torqued bolts, dust-free spindles. Skipping maintenance because "it's still working" is the fastest way to guarantee it won't keep working.

When the spindle seizes mid-job, the stepper belts snap, or the bearings scream louder than your shop radio, you'll wish you'd done that little bit of weekly TLC.

Remember: maintenance now is like exercising; the benefits aren't immediate, but over time, your machine, and body, will reward you with reliability and longevity.

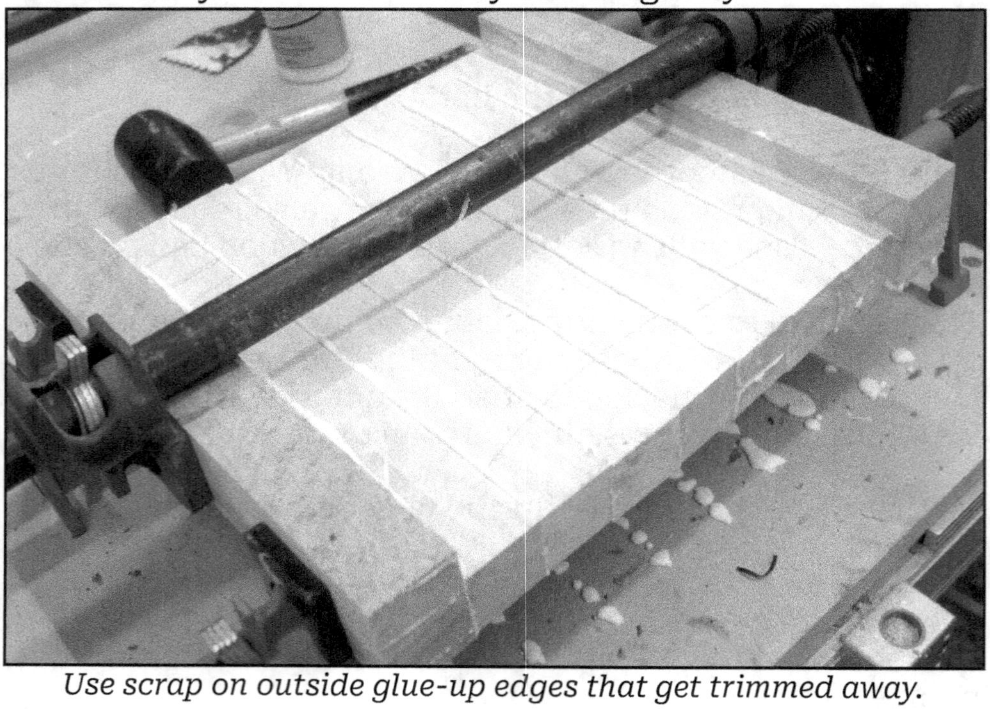

Use scrap on outside glue-up edges that get trimmed away.

Veneers and stains are easily removed by routing.

1.4 Lessons I've Learned About Routers

Respect the Router.

CNC routers are incredible machines, opening up worlds of precision and creativity. But they're still powerful tools, and powerful tools deserve respect.

Falling for myths or ignoring safety practices can turn your dream project into a disaster, or lead to an awkward call to tech support and unexpected downtime at the worst possible moment, right in the middle of a unicorn job.

So wear your PPE, keep an eye on the machine, respect your tooling, and treat maintenance as part of the craft.

Remember, it's not just about finishing the job, it's about staying within tolerances and working safely, without pulling a Simon by accidentally setting your shop on fire.

Because in the end, the best part of CNC routing isn't just watching your design come to life, it's being around to enjoy it when the job is done.

1.5 How to Read Tool Wear Like a Detective

CNC bits don't quit loudly, they tend to fade before quietly snapping. Your CNC won't even flinch! If your cuts are getting rougher than your Monday mornings, your bit's probably on its way out. I've shattered half-inch bits at *CNCROi.com*, so don't assume thickness means invincibility.

Why Your Cuts Suddenly Suck

Tool wear is CNC's version of a midlife crisis: the bit dulls, cuts go bad, and weird noises start with material vibrating off the bed. Before long, you're dealing with burn marks, tear-out, melted plastic, snapped tools, and a serious drop in precision. It stresses the spindle, ruins parts, and turns a clean job into chaos.

Ignoring it wastes material, time, and patience, and nothing fuels machinist rage like a ruined part from a dull bit. If your machine sounds like a blender full of nails, it's not tough, it's in trouble.

Keeping tools sharp means cleaner cuts, happier machines, fewer delays, and less swearing.

Standoffs provide excellent support for painting and staining.

The 5 Faces of a Dying Bit (CSI: Toolpath Edition)

Edge Dulling often shows up as a glossy edge, burn marks on wood, or just an overall sad-looking surface finish. If you're hearing a high-pitched whining sound, don't worry, it might not be you. The fix is simple: retire the bit to your carbide jar for recycling. Don't worry, you hopefully made money during its tenure. Remember, your bit doesn't owe you anything, especially after five brutal jobs cutting MDF.

Micro-Chipping shows up as tiny nicks along the cutting edge, they can be small, but devastating. It feels like betrayal, and honestly, it kind of is. The usual suspects? Aggressive feeds, rogue knots, or that surprise screw hiding in your material. The fix: slow your roll, literally. Take shallower passes, adjust feed per tooth (FPT) and ease into cuts more gently. And remember: sharpening only makes sense if the part is worth more than your weekend.

Built-Up Edge (BUE) happens when aluminum cakes onto your bit like frosting but without any of the sweetness. What's really going on? You're unintentionally welding metal to your tool, and trust me, it's not into that. The fix: switch to a bit designed for aluminum, add mist coolant, and lower your RPM to keep temps in check. Aluminum isn't forgiving, it's clingy and jealous. Use lube.

Flute Gumming looks like your bit took a dive into melted marshmallow. Plastic sticks to the flutes, chips clog instantly, and smoke starts rolling out like you're summoning spirits from the spindle. The cure? Increase your feed rate, drop the RPM, and maybe say a few kind words to your overworked shop vac. If it smells like burning crayons, your bit's not cutting, it's suffering.

Coating Erosion shows up when bare metal starts peeking through what used to be your bit's high-tech armor. You'll notice more friction, rising heat, and that creeping realization that you're one pass away from scrapping a part. The fix? You're done. That coating was the bit's protection, it's naked now. Be kind and retire it before things get ugly, as things tend to get ugly fast spinning at 18 000 RPM.

Your Chips are Tattletails

Chips don't lie, they're the CNC machine's way of telling you exactly how things are going, whether you like it or not. The shape, color, and texture of the chips coming off your tool can reveal everything from a perfect cut to a slow-motion disaster.

Learning to read chip style is like reading tea leaves: a quick glance can warn you when something's off before it turns into a broken tool or a scrapped part.

Chip Style	Interpretation	What to Check
Powdery dust	*You're rubbing, not cutting.*	*Increase feed rate or reduce spindle speed.*
Dark, smoky bits	*Too hot. Lower the RPM and walk away slowly.*	*Check RPM, coolant flow, and tool sharpness.*
Sticky blobs	*I melted. Hope you weren't attached to that part.*	*Reduce speed, improve chip evacuation, cool more.*
Curly & dry	*This is what success looks like. Don't touch anything!*	*Keep feeds, speeds, and tooling as-is.*

Machine Mood Swings

Your CNC has opinions, and it expresses them loudly. A good bit gives you a steady hum, the sound of trust. A worn bit adds a high-pitched whine. A dead bit? That screeching you hear is the sound of a pending lawsuit... or dead quiet (it broke and you are now cutting air with what's left).

Other clues are just as telling. If your finish looks furry, you're dealing with a dull edge.

If the collet feels gritty, clean it before it files for divorce. And if you smell toast but haven't made any, your bit is overheating, use your emergency stop if you see smoke.

Preventing Premature Bit Death (Bitmortem)

Let's make your tools last longer than your patience. Resin and debris buildup on the flutes traps heat, accelerates wear, and wrecks surface finish, especially during long jobs or when cutting gummy materials or laminations. Scrape it off regularly, ideally between jobs or even mid-run if buildup gets heavy. If your bit looks like it's been dipped in tree sap, don't just sigh and keep cutting. Clean it. A bit that's running hot and dirty is living on borrowed time.

Like a CNC laser lens, clean cuts better than dirty and produces better results. A few minutes of maintenance can save you from trashed material, wasted hours, and a fried spindle. Preventative care keeps machines happy.

The CNC Bit Retirement Plan

Instead of letting worn-out bits float around the shop like CNC zombies, set up a "tool retirement box" to keep them out of rotation and out of trouble.

Send your bits off to the big shop in the sky with a proper send-off, metal recyclers pay top dollar for carbide!

You can get excellent pricing on rough lumber.

Use the Right Bit

Don't use your MDF upcut bit on aluminum, that's not bold, it's just suicidal. For plastics, stick with single-flute or O-flute bits. They're designed for clean, cool cuts, and using the right tool saves you from meltdowns, both literal and emotional.

When to Let Go

If your tool is smoking, louder than your compressor, or carving modern art instead of clean parts, it's time to retire that bit, or check your spindle (time for a rebuild?). Try to end a bits career one job early, not one job too late.

Rough Tool Life Estimates (a.k.a. Your Bit's Horoscope)

Material	Tool Life	Reality Check
ABS	*5–8 hrs*	*Melts easily, keep it cool or it'll gum up everything.*
Acrylic	*6–10 hrs*	*Unless you melt it first.*
Aluminum	*2–4 hrs*	*Requires lube and prayers.*
Carbon Fiber	*1–2 hrs*	*It cuts your bit, not the other way around.*
Foam	*10+ hrs*	*Unless you go full BBQ mode.*
HDPE	*6–10 hrs*	*Cuts clean but loves to melt if you push it too fast.*
MDF	*2–6 hrs*	*Glue and dust will eat your edge alive.*
Plywood	*3–6 hrs*	*Each glue layer = microscopic betrayal.*
Polycarbonate	*4–7 hrs*	*Tough, clear, and a pain when it gets hot, don't rush it.*
PVC	*4–6 hrs*	*Cuts okay, but smells like regret and sticks like melted gum.*

Trust Your Gut, and Your Collet

If your machine starts sounding off, don't point fingers at the G-code. Don't blame the stock. Blame the shiny little liar spinning at 24 000 RPM. That tool is trying to tell you something and ignoring it usually ends with a whole lot of sanding, something you really want to avoid having to do.

Learn to listen. A subtle change in sound often means a not-so-subtle problem. Don't worry, you can hear it through your PPE, so don't use this as an excuse to not wear it.

Pick the Right Machine for the Job

Not all CNCs are created equal, and using the wrong one can turn a simple job into a slow-motion tool massacre. If you're frequently cutting aluminum, it might be time to stop torturing your router and invest in a CNC mill. Routers can handle occasional aluminum jobs, but extended runs bring heat, vibration, and premature tool wear that will slowly destroy your spindle and your patience. Mills are purpose-built for metals as they are generally heavier, more rigid, and designed to take the punishment that Inconel dishes out. Save your router for just about everything but metal or hard plastics; use the CNC mill for the rest.

If your workflow is packed with acrylic projects, consider a laser cutter. While a CNC router can cut acrylic, it often leaves rough edges, melting, and that telltale cloudy finish. A CO_2 laser, on the other hand, delivers sharp, flame-polished edges with no contact and no tool wear. It's faster, cleaner, and often requires less post-processing, especially for signage, enclosures, or decorative work.

Materials like wood, MDF, foam, and composites are where CNC routers shine. Lightweight and quick, routers are ideal for these materials, but even then, it's important to match the machine size and rigidity to the job. Cutting oversized hardwood slabs with a trim router mounted on rails is a shortcut to chatter, inconsistency, and sad-looking joints.

Ultimately, using the right machine for the right material isn't about being fancy, it's about working smarter.

☐ Let the Chips Fly, It's Q&A Time!

1.6 How does router bit geometry affect edge quality?

Router bit geometry isn't just nerdy trivia, it's the secret sauce behind clean edges, accurate cuts, and keeping your project from looking like it got gnawed on by a squirrel.

The sharpness of the cutting edge, flute angle, rake, and helix direction (upcut, downcut, and the mighty compression) all work together like a tiny high-speed orchestra, with the material you're cutting as the headliner.

Sharp, polished carbide edges are your best friend for smooth cuts and minimal tear-out, especially in wood and plastics. They're like the sushi knives of CNC, precise, clean, and oddly satisfying. Flute geometry also matters.

Straight flutes are great for soft stuff like MDF, chill and low-drama. **Spiral flutes** evacuate chips better and reduce tool pressure, making them ideal for hardwoods and composites. With my ShopBot Desktop cutting plywood when I first started, this was literally all I used!

Downcut spirals press fibers down for a crisp top edge, perfect when looks count.

Upcut spirals pull chips up fast, but can fray the top surface.

Compression bits do both, clean edges on both sides, perfect for plywood and showing off.

Fewer flutes = bigger chips = better roughing.
More flutes = smoother cuts = better finishing.

1.7 What's the effect of feed rate on chip evacuation?

Feed rate, which is how fast the tool moves through material, is tightly linked to chip evacuation efficiency and cutting performance.

If your feed rate is too slow, the bit isn't cutting, it's rubbing. This creates excessive heat, which can melt plastics, burn wood, rapidly dull your bit, and unintentionally replicate the ancient art of fire-starting and if you have a vacuum table, you've added gas to the mix!

Instead of clean chips, you get dusty, clingy debris that clogs your flutes and sticks around like a bad houseguest. Heat builds up, chips recut, and suddenly your "precision job" looks like a campfire accident.

On the flip side, if your feed rate is too fast, you risk throwing chunky chips the size of granola, clogging your dust collection, and pushing your machine past its torque limit. The result? Chatter, tear-out, and maybe even snapping a bit.

The sweet spot lies in chip load, that magical balance between feed rate, spindle speed (RPM), and bit geometry.

Aim for chips that are not too fine, not too thick, just right.

Dial it in, and you'll get cleaner cuts, cooler tools, and faster jobs. Ignore it, and you're just causing your own problems.

Do not fear long bits. More length means more cutting depth.

1.8 How do you calculate optimal stepover for 3D surfacing?

Optimal stepover, which is the distance between adjacent tool passes in 3D machining, is the unsung hero of surface finish and machining time. Get it wrong, and you'll either be sanding for hours or watching your machine run until it qualifies for a pension.

Too large a stepover, and you'll end up with visible scallop marks, as if your part caught a case of CNC chickenpox. Too small, and you get a silky-smooth surface... along with a dramatically and exponentially longer carving times.

A good rule of thumb? Set stepover between 40 - 60% of the tool diameter. So, for a a 3/8 inch (10 mm) ball nose, with about a 3/16 inch (5 mm) stepover hits that sweet spot between detail and speed. Smaller stepover = smoother finish, but longer wait. Bigger stepover = faster cut, but more sanding. Choose your pain.

Tool shape matters too as ball nose vs. flat vs. tapered all leave different surface textures. Material hardness and finish expectations also play a role. For fine detail work (or if you're allergic to sandpaper), stepover might shrink to 10–20%, so feel free to test these on scrap before a project.

Most CAM software even gives you a preview of scallop height, so you can obsess over tiny ridges before wasting six hours machining them. They aren't always accurate.

In the end, stepover is all about balance, between beauty and efficiency, art and productivity, sanding and muscle.

1.9 What's the role of toolpath simulation before running a job?

Toolpath simulation is a critical step in CNC routing that allows the operator, yes, probably YOU, to preview how the machine will cut a part before any material is touched.

This virtual dry-run helps catch problems that could waste material, damage tools, or even harm the machine.

A good simulation doesn't just look cool, it shows the full tool motion: rapids, entry points, cut depths, and directions. Think of it as a dress rehearsal for your bit, minus the applause. It ensures your toolpaths actually follow the design, not some rogue alternate reality version of it. I always run simulations at *CNCROi.com* and provide screenshots to the customer for project approval.

Simulations help catch the scary stuff, like a long tool crashing into your clamp while you're halfway through a beautiful cut. They also expose time-wasters like unnecessary retractions or your machine aimlessly cutting air like it's practicing karate with a ghost. Why go 2 inches above the surface for bit retractions when 1/2 inch (12 mm) does the same job, and saves time!

You'll also catch overcuts, missed spots, and weird start points, often caused by CAD-to-CAM translation gremlins or post-processing hiccups. On complex jobs (especially multi-sided ones), simulation helps confirm that your workpiece is aligned correctly and not, say, upside-down.

Sawdust happens. A well set machine is built to work through it.

1.10 How do spoilboards improve routing accuracy?

A spoilboard is the sacrificial surface bolted to your CNC table, a noble layer whose job is to get chewed up so your machine doesn't have to. It protects the table and your bits during full-depth or through-material cuts, and it quietly plays a big role in routing accuracy.

First, it lets you cut slightly into the board itself without panic, aim for 0.01 in (0.25 mm). This ensures your parts actually separate from the workpiece, or that your pockets hit full depth.

Second, a freshly surfaced spoilboard gives you a perfectly flat, level reference plane. That's critical for consistent depth across the entire bed, especially on large-format machines where the table might have minor warps, dips, or the CNC equivalent of a bad back.

Third, spoilboards actually help with vacuum hold-down systems. Porous MDF, for example, allows vacuum suction to pass through the board itself, which is perfect for securing thin or small parts that would otherwise fly off.

They're also your workholding wingman. You can screw or clamp parts directly into the spoilboard without flinching because it's sacrificial, and that's its whole job. No guilt, no bed damage, just solid hold-down. You can fill the resultant screww holes with glue afterwards.

And by resurfacing (a.k.a. flycutting) the spoilboard now and then, you keep it flat and parallel to the spindle. That means cleaner cuts and consistent depths.

1.11 What vacuum hold-down methods work best for thin materials?

Bed vacuum hold-down is one of the best ways to secure thin sheets like plywood, plastic, or even aluminum.

But like most good things in CNC life, it requires careful setup and a bit of strategy for optimal results regardless of what you are putting under the gantry.

Porous spoilboards (like MDF) paired with a high-flow vacuum pump can spread suction evenly across the entire surface, gently anchoring lightweight materials like they're being hugged by air pressure. It's quiet, clean, and avoids harsher alternatives. Stronger the vacuum, the easier life is.

For non-porous materials or tiny parts that love to shift, gasketed vacuum zones or pods step in. These systems seal off specific areas, concentrating suction where it matters and preventing vacuum loss in the zones your material doesn't cover because wasted suction is best used directed towards the part you are cutting.

Need more hold? Masking tape on the underside of your material can help seal micro-gaps, especially for thin plastics that warp if you so much as look at them wrong.

A sacrificial layer, like paper or thin plastic film, can also help increase surface contact and boost hold. You see me doing this in a lot of *CNCROi.com* videos using acrylic and MDF off-cuts. I've also been known to use a glue gun too!

Pre-visualization maximizes material use and minimizes waste.

1.12 How does gantry stiffness influence cutting precision?

Gantry stiffness, a fancy way of saying how stubborn your router's horizontal beam is, plays a huge role in cutting precision, especially when you're slicing through tougher materials, digging deeper, or speeding things up.

A stiff gantry holds its ground when cutting forces try to push it around. If it's too wobbly, it flexes like a yoga instructor, throwing off your toolpath, leaving chatter marks, and generally making your cut look like it was done during an earthquake. Even a tiny bend can knock your cutter off course enough to mess up tight tolerances or leave a surface rough enough to sand with your eyeballs.

Stiffness comes from using solid materials (like steel or beefy engineered aluminum), smart design (think big, reinforced beams), and strong structural supports (quality rails, carriages, and bearings). Lightweight or flimsy gantries might be okay for carving soft stuff like foam or balsa, but they'll cry uncle on aluminum or hardwood. And don't forget, even the stiffest gantry is only as good as the base it's bolted to. If your machine's frame wobbles, it's like putting racing tires on a shopping cart.

A tough, well-built gantry gives you cleaner cuts, longer tool life, and lets you work faster, all without sacrificing accuracy. Essentially, the heavier your router (or CNC in general), the more rigid it is overall.

1.13 What are the trade-offs between collet types (ER vs. proprietary)?

Collets are the unsung heroes of your CNC router or mill setup; they are the little components that grip your cutting tools and keep them from flying off mid-job (which, trust me, is not ideal). The type of collet you use affects how flexible, precise, and user-friendly your tool changes are.

ER collets (like ER11, ER16, ER32) are the Swiss Army knives of the collet world.

They're industry-standard, super versatile, and can snugly hold tools across a range of diameters, no need to hunt down a new collet every time you switch bit sizes. Whether you're running delicate engraving tools or chunky end mills, ER collets have you covered.

They provide excellent concentricity (low runout), are widely available, and allow shops to switch tools easily between machines or brands. They're also relatively affordable and compatible with many after-market toolholders.

However, they may not achieve the ultra-high precision or rigidity of some proprietary collet systems specifically designed for certain CNC routers or spindles.

Proprietary collets, on the other hand, may offer tighter integration with the machine, optimized fit, or higher clamping forces, but they lock the user into a specific manufacturer's ecosystem, making replacements or upgrades more expensive and harder to source.

Plastic sheets provide durable, reusable vacuum table blocking.

1.14 How do you balance pass depth and tool life?

Balancing pass depth and tool life in CNC routing is like trying to slice a cake fast without wrecking the knife, or the cake. Go too deep, and your tool screams, your spindle sweats, and everything starts shaking like it's auditioning for an 80s dance show.

Pass depth, in plain terms, is how far your tool dives into the material each time. If it cannonballs in too deep, expect heat, wobble, premature tool death, and parts that look like they were cut by a blindfolded raccoon.

But if you tiptoe in too lightly, your job turns into a marathon of baby steps. That means more passes, longer runtimes, and your tool slowly dying of boredom and overuse. Also, consider your Z-axis speed. Plunging too fast can stress the tool and spindle, while too slow increases runtime without improving quality.

Of all the problems you can have with a CNC router, this is by far the easiest one to fix as you directly define this in your software.

Rough first, then refine with smaller tools for clean, precise results.

Usually, pass depth is set to about half to one times the tool's diameter for roughing, basically, the CNC version of "go big or go home," but with some common sense. The exact number depends on how tough the material is, how beefy your tool is, and whether your machine's got the muscle or just skips leg day.

For finishing passes, you go gentle, think spa treatment, not sledgehammer, so the surface ends up silky smooth, not looking like a cheese grater had a meltdown.

Using sharp, high-quality cutters (not those bargain-bin ones) and a nice dust extraction setup keeps them happy longer. Plus, if you tune into the cut like it's a symphony: listening for weird sounds, checking chip size, and watching for burn marks. You can adjust pass depth like the CNC whisperer we all know you are.

1.15 What's the benefit of mist coolant systems for aluminum routing?

Routing aluminum can be a hot mess, literally. The metal's softness makes it stick to cutting tools, forming built-up edge (BUE), wrecking surface finish, and chewing through tools like popcorn.

Enter the mist coolant system: a fine spray of lubricant and compressed air that keeps things cool and clean. The lube cuts friction, so aluminum doesn't weld itself to the bit, while the air blasts chips away like a tiny tornado.

Unlike the flood coolants common in heavy-duty CNC mills, mist systems are enclosure-friendly and easy to add to your existing router setup.

The payoff? Cleaner edges, tighter tolerances, and tools that live to cut another day.

For shops doing regular aluminum work, adding a mist coolant system can transform their router from a wood-focused machine into a versatile hybrid capable of reliable light metal machining. **You MUST have a solid gantry though, otherwise, tolerances are out the window.**

1.16 How does router vibration affect cut accuracy?

Router vibration is like that one guy at a concert who claps off-beat, absolutely killing the vibe. In CNC routing, vibration is one of the top enemies of precision and surface quality but the good part about it is that you can hear it.

It can sneak in from multiple directions: mechanical gremlins like loose bearings or out-of-balance spindles, operational missteps like wonky feed rates or spindle speeds, or even material quirks like dense grain or surprise hard spots like a piece of iron in your MDF.

When I switched from my ShopBot PRS Alpha with a manual tool changer to an HSK ATC on a Thermwood MD42, it was like trading a rowboat for a stealth bomber.

The difference in resonance control and vibrational accuracy when prototyping molds was night and day. That upgrade didn't just help, it practically gave the toolpath a hug and told it everything was going to be okay.

Vibration isn't just annoying, it's a precision killer. It causes chatter marks, dimensional errors, and sends your favorite end mills to an early grave. Even slight tremors can throw off tight-tolerance features you spent hours dialing in. You can also route faster with less vibrations.

Your CNC needs to be rock-solid: rigid gantry, stable frame, and a spindle mount that doesn't flinch. Use sharp, balanced tools, high-quality collets, and clamp them tighter than a toddler clutches candy. Optimize feeds and speeds, too fast, and your bit wails; too slow, and the material pushes back.

Secure your workpiece like it's plotting an escape, no movement allowed. Maintain your machine with bolt checks, clean rails, and a keen ear for weird noises (unless you're composing CNC jazz).

Tame the shake and you'll get better parts, faster runs, and happier tooling. Tame the jitters in your machine, and it performs like a champion athlete.

1.17 What maintenance routine maximizes spindle life?

The spindle is the heart of your CNC router, treat it right or prepare for heartbreak and hefty repair bills.

Daily, clean collets and toolholders like you're prepping for surgery, not sanding drywall. Don't overtighten unless you enjoy snapped collets and angry bearings.

Weekly, check the cooling system, air or water, just make sure it's actually cooling. Overheating is a top reason spindles tap out early.

Monthly, inspect bearings for play or roughness and check runout to make sure the spindle's not wobbling like it just woke up. Clean or swap filters if your spindle has them.

Annually (or per the manual), it might be time for a pro checkup or bearing replacement. Spindles need some love too. You can often detect issues by sound long before the machine seizes.

Long bits make it harder for dust systems to clear chips.

1.18 How do you correct for tool deflection in long-reach cuts?

Tool deflection is what happens when your end mill decides it's had enough and starts bending like it's auditioning for a gymnastics team.

It's a common drama in deep-pocket or long-reach machining, especially when your tool's length-to-diameter ratio starts creeping past 3:1 or 4:1; basically, when your cutter is more noodle than tool.

To keep your cutter on the straight and narrow, start by minimizing stick-out. Only extend what you absolutely need. Shorter tools = less flex = happier parts.

Next, tweak your cutting strategy. Ease up on depth and width per pass, and bump up spindle speed a bit to reduce cutting pressure. Think finesse, not brute force.

Good CAM software adds a secret weapon: spring passes.

Oversized stock gives you valuable scrap for future builds.

These feather-light final cuts clean up what deflection left behind without bullying your toolpath.

And don't forget cutting strategy. Slowing your feed and switching to climb cutting (when it makes sense) helps redirect forces more kindly. Still not sure? Run a test on scrap first.

It's always better to ruin junk than your workpiece.

1.19 What are common safety hazards specific to CNC routing?

CNC routing: where wood meets whirring doom, and safety goggles are the difference between a productive day and earning your pirate credentials early.

First up, the spinning spindle of death. It moves fast enough to qualify for a speeding ticket and doesn't care if your fingers are in the way. Bypass the guards and it's basically inviting it to bite.

Next, dust and chips. These machines throw off more sawdust than a beaver hopped up on meth. Fine dust tries to sneak into your lungs like a ninja, while larger chips launch themselves across the room like sharp popcorn.

Then there's the noise, loud enough to make a rock concert seem like meditation. Forget hearing protection and you'll be communicating in charades by retirement.

Electrical hazards? These machines guzzle power like a thirsty vampire, and one frayed wire could deliver a three-phase shock that feels like touching the Sun.

And let's not forget kickback: when the workpiece decides it's had enough and tries to escape at Mach 3. Best case, it hits a wall. Worst case, launching straight at your head!

That's why smart operators wear the holy trinity of PPE: safety glasses (because eyes are useful), earplugs (for future conversations), and dust masks (lungs: keep 'em).

1.20 How does climb cutting differ from conventional cutting?

Climb cutting and conventional cutting sound like extreme sports, but they're really just about which way your CNC tool is spinning when it meets the material, like a very sharp, very fast handshake.

In climb cutting, the cutter spins in the same direction as the material feed, so it's basically chasing the workpiece like a dog after a squirrel. This gives you smoother cuts, less tear-out, and happier tools. But there's a catch: it also tries to yank the workpiece into the machine like it's claiming a sacrifice. If your material isn't locked down tighter than a toddler with candy, it's going airbone.

Conventional cutting, on the other hand, is more of a head-on collision. The tool spins against the material's movement, scraping and gnawing at it like it's trying to pick a fight. The edge might be a little rougher, but the workpiece stays put, and your blood pressure stays lower.

Most pros use a combo platter: conventional cuts to rough things out (and keep stuff from flying off the table), then climb cuts to smooth the edges.

Knowing which method to use, and when, is the key to keeping your project clean, your machine happy, and your eyebrows intact.

1.21 How do you optimize nesting layouts for minimal waste?

Nesting is how you cram as many parts as possible onto a sheet without turning it into a game of Material Tetris.

Good nesting starts with smart software.

Advanced CAM programs do the heavy lifting, automatically arranging parts like an overachieving jigsaw master, taking into account grain direction, cut allowances, and your material's hopes and dreams.

Operators should group similar parts together, not just for friendship, but to cut down on tool changes and keep the machine humming. When working with plywood or MDF, following the grain isn't just for hipsters, it actually makes your parts stronger and prettier.

Don't forget kerf, that's the width of the cut, not a fantasy creature. If you pack parts too tightly, the router might start snacking on the edges of your dreams. Leave a little breathing room.

Want to keep tiny parts from launching into orbit? Add tabs or bridges to hold them in place. Otherwise, your dust collector might decide it's hungry and eat them.

Rotating oddly shaped parts and using offcuts from past jobs is nesting on hard mode, but it saves material and feels oddly satisfying.

If you're doing a lot of cutting, take time to review and tweak your nesting game after each production run.

Why does everyone ask for Shaker style cabinet doors?

1.22 What software post-processors work best for complex joinery?

Post-processors are like translators for your CNC machine, they take your carefully crafted CAM toolpaths and turn them into machine-friendly G-code, so your router doesn't just sit there blinking in confusion.

When it comes to complex joinery, think dovetails, box joints, or those fancy interlocking finger joints that make wood nerds weep with joy, your post-processor has to be on its A-game. Don't fear experimentation on this front.

That means handling multi-axis moves, tool changes, and telling the machine exactly when to start, stop, and not crash on you in the middle of a production run.

Popular CAM programs like Vectric Aspire (I use VCarve Pro at *CNCROi.com*), Fusion 360, and RhinoCAM offer solid post-processors tailored for controllers like Mach3, or the fantastic ShopBot (my first router).

Use common plumbing tubing to build your own vacuum system.

If your joinery is extra fancy or unusual, you may need a custom post, like a CNC cheat code, to add pauses, offsets, or other magic for perfect cuts. Honestly, in over two decades of CNC, I've never had to resort to this for any client projects I've completed at *CNCROi.com*.

Running 4- or 5-axis? Congrats, now get a post-processor that can handle your machine's interpretive dance without freaking out mid-dovetail.

Always test first with simulations or scrap. Nothing says "bad day" like turning premium stock into scrap bin filler.

The right post-processor means smoother workflow, less code tweaking, and joints that actually line up the first time.

1.23 How do you set up double-sided machining accurately?

Double-sided machining is like flipping a pancake, but if you're off by a hair, you're serving up expensive scrap.

Use dowel pins or alignment holes to keep things lined up when flipping. Drill matching holes in the part and spoilboard so everything snaps back into place.

Fixture plates or vacuum jigs also work, as long as your part stays locked down like it owes you money.

In your CAM setup, use the same zero point for both sides: whether that's the center, a corner, or a dowel hole. Don't forget to factor in material thickness and any Z-axis offsets, or your "perfect match" will be more of a mismatch.

And always test on scrap. Yes, I've said it a million times, and I'll say it a million more. No matter what CNC platform you're planning on using, this is the only way to get consistent, disaster-free results.

Get it right, and double-sided machining delivers crisp, symmetrical parts, from custom enclosures to slick 3D carvings to mechanical bits that actually fit together.

1.24 What are the best materials for jigs in repetitive CNC jobs?

Running the same CNC job over and over? You need a good jig, basically the CNC equivalent of a precision-driven sidekick powered by curiosity.

MDF is the go-to for wood and composite work. It's cheap, stable, easy to cut, and holds vacuum like it's clinging to life. Perfect for simple templates or vacuum jigs that don't need to last forever, just long enough to get the job done without little fuss.

Need something tougher? Phenolic boards are like MDF's older, buffer cousin. They resist wear, stay accurate, and won't flinch under pressure, ideal when your jig sees more action than your weekend plans.

Aluminum is the VIP of jig materials: strong, precise, and basically immortal. Great for long-term use or when "close enough" isn't close enough. Downside? It's pricier and takes longer to machine, so bring snacks. *The "harder" the jig material, the more mechanical wins over vacuum holding.*

If you just need a quick, no-fuss jig, acrylic or HDPE plastics are lightweight, easy to cut, and forgiving when tools bump into them, unlike your ego after a ruined part.

In the end, the best jig material depends on your budget, how long it needs to survive, and how precise your job is. But a good jig makes life easier, parts repeatable, and your CNC workflow smoother than a fresh sheet of Baltic Birch.

1.25 How can CNC routers be used to create hybrid wood-metal parts?

CNC routers aren't just for woodworking, they're also great at helping wood and metal become best friends (with benefits like strength, style, and precision).

Need wood to hold metal? Your router can carve out perfect pockets, slots, and mortises for brackets, inserts, or fasteners, tight fits every time, no guesswork, no wobbles.

MDF is highly versatile when machined on a CNC router.

Threaded inserts in CNC-cut holes turn wooden parts into mechanical champs, not just pretty faces. They add serious strength, allowing wood to handle bolts, screws, and repeat assembly without turning into splinters and regret.

Want to get fancy? Routers can cut clean channels for metal inlays, think aluminum or brass bling in your Black Walnut masterpiece. Logos, patterns, or even subtle highlights can take a project from "nice" to "is that custom?"

In some next-level shops, CNC routers equipped with mist coolant systems can even machine light metals like aluminum directly, letting one machine tackle both materials without breaking a sweat (or your budget).

Just remember: wood and metal don't behave the same. Wood swells when it's humid; metal shrinks when it's cold. Throw in a seasonal shift and your project could start moving like it's alive. Design your joints with a little wiggle room and think about how the materials will age together, kind of like relationship counseling, but with fasteners.

1.26 How do you adjust feeds and speeds for hardwood vs. softwood?

Feeds and speeds are the magical combo of how fast your spindle spins and how quickly your tool moves, need to be dialed in carefully, especially when choosing between hardwood and softwood, or natural and engineered materials. Get it wrong, and you're either scorching your project or launching splinters into the stratosphere.

Hardwoods are the tough guys: dense, stubborn, and prone to problems if you push too hard. Go slow on the feed rate and ease off the RPMs a bit, or you'll burn the wood, gum up your bit with sticky resin, and make your cutter question its life choices.

Softwoods are more laid-back, they cut easily and actually like higher speeds and faster feeds. Crank it up and you'll get cleaner edges and fewer fluffy tear-outs that make your finish look like a bad haircut.

No matter the wood, sharp bits are non-negotiable. Compression or downcut spirals keep laminated hardwoods looking sharp, while upcut spirals in softwood clear chips like they're on a mission.

Don't forget chip load, too little and you're just polishing the wood while your bit cries quietly; too much and you're one snap away from replacing tools... again.

CAM software gives decent starting points, but real CNC pros know the truth: do test cuts, listen to the machine, and adjust until it sounds like it's purring, not screaming.

1.27 What are the benefits of upcut vs. downcut router bits?

Upcut and downcut bits have different personalities, knowing when to use each is the secret to clean cuts.

Upcut bits yank chips upward, great for clearing debris and keeping things cool in deep cuts through wood or plastic.

They treat the top surface like it's a trampoline, bouncing around chips and splinters everywhere, not ideal for anything meant to be seen.

Downcut bits do the opposite: they press fibers down, leaving the top looking sharp and splinter-free, perfect for sign faces or cabinet panels. The catch? They mash chips into the cut, which can cause heat and clogging if you're not feeding them right. Bits also don't last as long from heat.

Compression bits are the overachievers. They pull up from the bottom and push down from the top, giving you clean edges on both sides. Ideal for laminated boards or double-faced plywood where both faces matter.

Choose the right bit based on your material and cut, or prepare for crispy edges, sad finishes, and tool tantrums.

1.28 How does stepper vs. servo motor choice affect router performance?

The motor type powering your CNC router's axes can make or break your machine's performance, precision, and budget. I've had CNC machines with both motor types. You get what you pay for but the route you take depends more on the projects you decide to take on than anything else.

Stepper motors are the budget-friendly, no-frills option. They move in little steps, don't need feedback, and get the job done for most entry- to mid-level machines. But push them too hard or fast, and they might miss steps, overheat, and start sounding like a dying printer.

Servo motors are the fancy ones, they use feedback loops and encoders to know exactly where they are at all times, like the overachiever in class. They're smoother, faster, quieter, and more powerful under load. Great for industrial machines, or anyone who hates redoing cuts.

The downside? Servos cost more, not just to buy, but to set up and maintain. But if precision and speed are critical, they're worth the splurge. If not, steppers will do just fine... with a little encouragement.

1.29 What dust collection systems are most effective for large routers?

Effective dust collection is the unsung hero of CNC routing. It keeps the air clean, your lungs happy, and your machine from choking on its own mess, especially when cutting wood, MDF, or composites that shed clingy dust.

For big routers, the best setups combine a high-powered extractor (think 3–5 HP... or more depending on need) with cyclone separators, smart hoods, and proper filters. The cyclone spins out chunky chips before they hit the filter, saving it from a cloggy death. Fine dust is then caught by HEPA or high-efficiency filters, essential for shops that don't want to double as dust clouds.

The router's dust shoe should hug the cutting zone like it means business, with flexible bristles to trap particles while moving with the bit. Bonus points for magnetic or quick-release shoes that make bit changes easy without flinging dust everywhere. Hose size matters too: big jobs need 4" (100 mm) or 6" (150 mm) hoses to keep up with the chaos.

Good system design avoids hose spaghetti. Keep in mind that fewer bends mean better airflow throughout.

Ground your system to prevent static buildup and surprise sparks (especially with dry materials).

Use blast gates or zone valves to direct suction where it's needed most, and size your ducts to keep dust flying toward the bin, not settling in the pipes.

Check filters, clean them, replace them, and make sure your hoses aren't secretly leaking like a shady garden hose. Consistent airflow is key.

Without solid dust collection, you risk more than just sneezing fits, as dust can wreck your cuts, gunk up your rails and motors, and in worst-case scenarios, turn your shop into a fire hazard with ambition. Avoid scenerios that might ignite the fine dust that CNC routers create.

CNC routers cut Formica well, but it is difficult to hold down.

The results look great, but plan for extra scrap.

1.30 How can sacrificial boards be used to improve workholding?

Sacrificial boards (aka spoilboards) do way more than just protect your CNC table from becoming a chew toy. They seriously boost workholding, especially for small parts or thin sheets that are too wiggly or fragile for clamps or vacuum alone. Screw or clamp your workpiece into the spoilboard, and voilà, secure cuts without troubles.

Using through-cuts? Go for it. The spoilboard takes the hit, not your machine bed. That's what it signed up for.

In vacuum setups, porous MDF spoilboards help spread suction evenly, keeping even tiny offcuts or paper-thin panels stuck in place like they owe you money.

Want extra precision? Mill custom pockets or negative shapes into the board to hold parts like a glove, instant custom jig. And when the surface starts looking like the moon, just flatten it with a fly cutter and it's good as new.

In short, spoilboards save your machine, secure your parts, and make setup way more flexible, so you can cut harder, smarter, and without wrecking stuff you'd rather not replace.

1.31 What is the impact of tool stick-out length on vibration?

Tool stick-out, how far your bit extends from the collet, might not sound exciting, but it's a big deal for cut quality, tool life, and whether your CNC ends up sounding like a blender full of bolts.

The longer the stick-out relative to the tool diameter, the more it likes to flex, wobble, and scream under pressure. That means ugly finishes, messed-up dimensions, and possible tool snap-age.

A good rule: keep stick-out to 3–4 times the tool diameter. So a 1/4" (6 mm) bit? Try to keep it within 3/4" - 1" (18–24 mm). Beyond that, you're entering noodle-territory.

Need more reach? Fine, just reduce your cut depth and width, slow your feed rate, or use long-reach or tapered bits designed to handle the extra stretch without crying.

Vibration gives itself away with chatter marks and that unmistakable "something's wrong" sound. Make sure your bits are properly seated, your collets are clean and in good shape, and nothing's wobbling that shouldn't be.

Keep your stick-out in check, and your cuts stay clean, accurate, and panic-free even in the deepest pockets.

1.32 How do you troubleshoot uneven cut depths across a large panel?

Uneven cut depths across a big panel? Time to play CNC detective. First suspect: the spoilboard. Over time, it can develop hills and valleys unless you resurface it regularly.

A quick fly-cut flattens things out and gives your bit a proper reference plane to work from.

Next, check if your spindle is tramming straight. If it's leaning like the Tower of Pisa, one side of your cut will dig in deeper than the other. Also inspect the machine bed itself, especially on large routers, where unsupported spans can flex like a trampoline under pressure.

Make sure your material is lying flat and clamped securely. A warped or bowed workpiece will laugh in the face of your Z-depth settings. Check underneath for splinters!

Then there's the tool setup: make sure the bit is seated properly, the collet is clean and undamaged, and nothing's wobbling where it shouldn't. Dirty or worn collets love to mess with your depth.

Lastly, double-check your software. If the Z-zero isn't consistent across tool changes, your machine might think it's cutting level when it's not.

Tackle each of these suspects, and your cuts will be even, smooth, and way less frustrating.

1.33 What role does gantry tramming play in cut accuracy?

Gantry tramming is the fine art of making sure your spindle is standing up straight while being perfectly perpendicular to the bed in both X and Y. Basically, you're making sure your machine isn't cutting with a tilt.

Proper tramming means your bit hits the material evenly across the whole work area: essential for planing, pocketing, and anything that needs smooth, consistent depths. If your spindle's off, even slightly, one side of the bit cuts deeper than the other, leaving annoying ridges and steps that scream "more sanding for you!"

Over large jobs, that tiny tilt adds up fast, wrecking part accuracy and turning clean finishes into lumpy disasters. Tramming usually involves a dial indicator or test bar in the spindle, swept across the table to check for unevenness. Then it's shim time, or maybe a little mechanical nudging, to bring things back into alignment.

Regular checks are a must, especially after crashes, moves, or tune-ups. A properly trammed machine means better parts, longer tool life, and fewer "why is this crooked?" moments.

1.34 How do you plan tool changes efficiently in multi-tool jobs?

Planning tool changes in multi-tool CNC jobs is key to keeping downtime low, errors rare, and efficient production. Make sure you group your toolpaths smartly. Don't switch bits like you're changing TV channels. Do all roughing with the big end mill first, then swap to smaller detail tools, and finish with the specialty stuff.

In your CAM software, sequence operations and assign tool numbers to match your machine's tool library or ATC.

For manual setups, label tools clearly, pre-measure lengths, and keep everything organized as random piles of bits lead to random mistakes.

Using preset tool length offsets? Excellent. That means you can swap tools without re-zeroing the Z-axis every time, saving precious minutes and avoiding the joyless ritual of poking around with feeler gauges after every cut. It keeps transitions quick, accurate, and far less likely to result in creative cursing.

Before you hit "go," double-check that every tool is sharp, properly loaded, and not secretly chipped, or worse, dull. A surprise mid-job tool failure is a one-way ticket to wasted material and lots more wasted time.

For complex jobs or production runs, always run a dry simulation or follow a checklist. It might feel overly cautious, but it's a small price to pay to avoid crashing a bit because path #4 accidentally ran with tool #7 instead of #8.

With just a bit of prep, tool changes become smooth and efficient. This leads to better parts, faster turnaround, and CNC operators who don't constantly look like they need a nap. Screw-ups are exhausting on the mind!

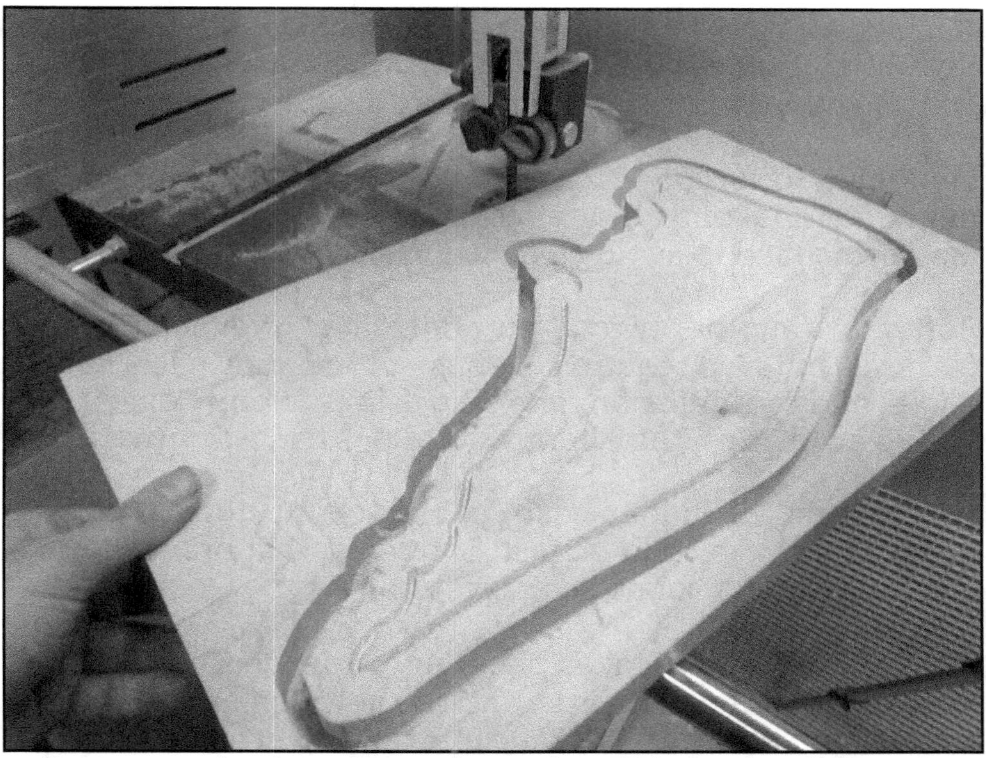

Custom cutting boards are always fun. This one is solid maple.

1.35 What are common pitfalls when retrofitting old routers with modern controls?

Retrofitting an old CNC router with modern controls is like giving a speed dial phone a smartphone brain. It can be awesome, but expect a few headaches.

I seriously considered retrofitting my old Thermwood CS43 that I bought at at auction a year earlier. It had only a 4-tool ATC, and I wanted more tools and a computer that didn't run like it was stuck in the '80s. But after weighing the cost and hassle, I just sold it (at a profit) and bought a wicked overkill Thermwood M42-55DT. It cost way more, but it was no longer obsolete. Sometimes it's better to cry once than be tech support for a dinosaur.

First up: compatibility. Older machines often have proprietary motor drivers, mystery encoders, or sensors that speak in extinct dialects. Getting them to talk to modern controllers usually means rewiring, reconfiguring, and a lot of colorful language.

Then there's mechanical wear. A shiny new brain won't help if your ball screws are sloppy, bearings are tired, or the gantry wobbles like a shopping cart on gravel.

Switching to servos? Great! But your old power supply might start gasping. Upgrades or isolation transformers may be needed to avoid electrical meltdowns.

Software's another twist. New controllers offer slick features, but also require new post-processors, a learning curve, and retraining anyone who thinks "floppy disk" is still relevant. Yes, the Thermwood CS43 had a floppy!

Don't forget the hidden costs: cables, connectors, safety interlocks as they add up fast. Suddenly, that brand-new machine doesn't sound so crazy.

A successful retrofit takes planning, patience, and a good grasp of both the old quirks and the new demands. Done right, it can turn a crusty relic into a lean, mean, cutting machine but look at other options as well.

1.36 CNC Router: Top 10 Takeaways

Router bit geometry matters: *Angle, sharpness, and shape affect cut precision, edge quality, and chip removal. The right bit minimizes tear-out, and improves finish.*

Feed rate affects chip evacuation: *Too slow creates dust and heat; too fast overwhelms the cutter. Balance feed rate and RPM for ideal chip load and cleaner cuts.*

Optimal stepover balances speed and finish: *Smaller stepovers give smoother results but take longer. Too wide leaves ridges, aim for 40–60% of tool diameter.*

Toolpath simulation prevents mistakes: *Simulate toolpaths to catch collisions, overcuts, and missed areas before wasting material or damaging tools.*

Spoilboards improve accuracy: *These sacrificial layers protect the machine bed, create a flat reference, and allow clean full-depth cuts.*

Vacuum hold-down secures thin stock: *Thin materials need strong or zoned suction to stay flat without clamps or distortion.*

Gantry stiffness ensures accuracy: *A rigid gantry resists cutting forces and prevents deflection, keeping parts dimensionally accurate.*

Collet type affects tool options: *ER collets fit a wider range of tools. Proprietary systems may offer precision but limit compatibility.*

Balance pass depth and tool life: *Deep cuts strain tools and spindles; shallow cuts waste time. Tune depth to your bit and material.*

Mist coolant helps with aluminum: *Mist cooling reduces heat, clears chips, and prevents aluminum from literally welding to the tool.*

⊗ 1.37 Section Quiz

This quiz is like a CNC router, if you don't follow the right path, you're gonna get cut!

1. What determines the width of the cut made by a router?
   A) Spindle speed
   B) Toolpath strategy
   C) Bit diameter
   D) Feed rate

2. Which material typically requires slower feed rates and shallower passes?
   A) MDF
   B) Hardwood
   C) Acrylic
   D) Foam

3. What is the main function of a spoilboard?
   A) Absorb vibrations
   B) Increase spindle power
   C) Provide a smooth surface finish
   D) Support workpiece and protect the machine bed

4. What does "stepover" refer to in 3D surfacing?
   A) The distance between each toolpath pass
   B) The maximum cut depth
   C) The spindle height above the material
   D) The thickness of the material

5. Which type of router bit is best for minimizing tear-out on plywood?
   A) Upcut spiral
   B) Downcut spiral
   C) Compression bit
   D) Ball nose

6. What does an ER collet do in a CNC router?
   A) Increase spindle RPM
   B) Hold the router bit securely
   C) Clamp the material to the bed
   D) Control coolant flow

7. What is the advantage of climb over conventional cutting?
   - A) Less heat buildup
   - B) Better chip evacuation
   - C) Cleaner edge finish
   - D) Faster spindle speed

8. What can cause tool deflection in a CNC router setup?
   - A) Short tool stick-out
   - B) Long tool stick-out with high cutting force
   - C) High spindle horsepower
   - D) Using a sharp bit

9. Why would you use a vacuum hold-down system?
   - A) To hold thin or flexible materials flat
   - B) To cool the bit
   - C) To improve spindle speed
   - D) To control dust collection

10. What's the role of a post-processor for CNCs?
   - A) Optimizes toolpath
   - B) Sharpens cutting tools
   - C) Increases spindle torque
   - D) Translates toolpaths into machine-readable G-code

11. What does the term "tramming" refer to?
   - A) Aligning the spindle perpendicular to the table
   - B) Leveling the gantry
   - C) Calibrating the servo motors
   - D) Resetting the machine origin

12. What's the main benefit of using mist coolant when cutting aluminum?
   - A) Better spindle life
   - B) Preventing corrosion
   - C) Reducing tool friction and clearing chips
   - D) Lowering machine vibration

13. Why are double-sided jigs important in CNC routing?
   - A) They improve cooling
   - B) They allow precise alignment of a part
   - C) They increase feed rate
   - D) They reduce bit wear

14. What effect does increasing the number of flutes on a router bit typically have?

A) Increases chip load
B) Reduces surface finish quality
C) Improves surface finish but reduces chip clearance
D) Lowers cutting temperature

15. What is one major risk of poor dust collection?

A) Decreased spindle power
B) Fire hazard from fine dust accumulation
C) Overheating the spoilboard
D) Lower vacuum pressure

16. What's the purpose of a ball nose end mill in CNC routing?

A) Cutting thin materials
B) Engraving letters
C) Drilling holes
D) Producing smooth 3D contours and surfaces

17. What kind of motor typically offers more precise control in a CNC router?

A) Stepper motor
B) Servo motor
C) DC brushed motor
D) AC induction motor

18. Why use a sacrificial board under your workpiece?

A) To increase spindle cooling
B) To improve nesting optimization
C) To protect the spoilboard and allow through-cuts
D) To reduce router noise

19. What toolpath strategy minimizes waste when cutting multiple parts?

A) Linear rastering
B) Pocketing
C) Random nesting
D) Optimized nesting

20. What is the purpose of toolpath simulation before running a job?

A) Check for collisions and errors in the cut sequence
B) Increase feed rate
C) Measure spindle deflection
D) Reduce spindle RPM

Routers are effective for surfacing and flattening uneven stock.

But sometimes another tool does the job better than the CNC.

✖ 1.38 Answer Key and Explanations

This answer key cuts deeper than a downcut spiral... but don't worry, these explanations are smoother than a ball nose on a jello sculpture.

1C – The bit diameter directly determines the width of the cut (kerf) made by the CNC router. A 1/4" bit cuts a 1/4" wide path. Spindle speed and feed rate affect quality, but not the cut width itself.

2B – Hardwoods are dense and tough, requiring slower feed rates and shallower passes to prevent burning, excessive wear on the bit, or chatter. Softer materials like foam or MDF can be cut faster and deeper.

3D – A spoilboard acts as a sacrificial surface under your material. It supports the workpiece during cutting and protects the CNC bed from being damaged during through-cuts.

4A – In 3D surfacing, "stepover" is the lateral distance between successive passes of the tool. A smaller stepover yields smoother surfaces but takes longer; a larger stepover finishes faster with rougher results.

5C – Compression bits have an upcut flute on the bottom and downcut flute on top. This design pulls fibers toward the center, reducing tear-out on both sides of plywood and laminated materials.

6B – An ER collet is a high-precision tool holder that grips the bit tightly in the spindle. This ensures stable, vibration-free cutting and accurate tool control.

7C – Climb cutting moves the tool in the same direction as the material feed, shearing fibers cleanly and resulting in a smoother finish, especially on wood. It can increase tool wear if not used carefully.

8B – Excessive tool overhang (stick-out) combined with strong cutting forces leads to tool deflection, where the bit bends slightly. This causes inaccurate cuts and can damage material or bits.

9A – Vacuum hold-down systems are especially effective for thin, flexible, or sheet materials, holding them flat and secure without mechanical clamps, which might interfere with toolpaths.

10D – The post-processor converts CAM software outputs into specific G-code formats that a CNC machine understands, tailored for each controller or machine model.

11A – Tramming ensures the spindle is perfectly perpendicular to the machine bed. This is crucial for even surface finishes during surfacing operations and accurate depth cuts.

12C – Mist coolant sprays a fine mist of lubricant/coolant, especially when cutting metals like aluminum. It helps reduce heat, clear chips, and prevent tool buildup, improving cut quality and tool life.

13B – Double-sided jigs help keep the workpiece accurately positioned when flipping it over, so machining on both sides aligns perfectly, critical for 2-sided operations.

14C – More flutes allow more cutting edges to contact the material, which typically improves finish quality. However, they leave less space for chip evacuation, which can cause clogging in softer materials.

15B – Poor dust collection allows fine dust to build up, especially from wood or MDF. This dust is highly combustible, creating a serious fire risk near hot tools or sparks.

16D – A ball nose end mill has a rounded tip that's ideal for 3D contouring and creating smooth curved surfaces like topographical models, sculptures, or molds.

17B – Servo motors offer closed-loop control, meaning they constantly adjust based on position feedback. This gives them higher precision and accuracy than stepper motors, which operate open-loop.

18C – A sacrificial board underneath your material ensures clean through-cuts without damaging the spoilboard or CNC bed. It also catches any minor tool over-travel.

19D – Optimized nesting arranges parts as efficiently as possible on a sheet to minimize material waste. It's essential in production settings where cost and efficiency matter.

20A – Toolpath simulation helps you visually verify the operation before running the CNC. It can detect collisions, tool errors, or missed steps, potentially saving time, material, and tools.

CNC waterjets can be messy to operate.

But they produce excellent blanks for further processing.

Section 2
The Waterjet Floor
Release the Kraken... With a Garnet Slurry

CNC waterjet cutting is one of the most fascinating and powerful technologies in modern manufacturing, it's like harnessing a tiny, ultra-fast river to slice through nearly any material you throw at it. Just like nature, but faster!

Using a high-pressure stream of water (generally mixed with abrasive particles), a CNC waterjet can cut metals, stone, glass, composites, plastics, rubber, and even food with precision that borders on magic.

Unlike lasers or plasma cutters, waterjets don't generate heat, meaning no heat-affected zones (HAZ) or warping, which makes them invaluable in industries where material integrity is critical.

At its core, a CNC waterjet uses digital design files to precisely guide the water stream along X, Y, and sometimes Z axes. This lets you cut incredibly detailed shapes, tight corners, and complex contours without worrying about tool wear, burn marks, or material hardness.

Whether you need to cut titanium, marble, bulletproof glass, or intricate gasket materials, the CNC waterjet doesn't blink, it just powers through. Slowly.

2.1 Why does industry love CNC waterjets?

First, its versatility is unmatched. Few other machines can handle such a broad range of materials and thicknesses, with a cutting capacity of up to 12 inches (≈ 300 mm).

Second, it's a cold-cutting process, meaning materials aren't heat-stressed or chemically altered, which is a big deal in aerospace, medical, and automotive applications where microstructure integrity matters.

Third, waterjets are eco-friendly compared to other cutting methods; they don't generate hazardous fumes, and the water can often be recycled within the system.

For companies working with high-value or exotic materials, a CNC waterjet is a game-changer. You can nest parts tightly to reduce scrap, cut stacked materials for higher throughput, and create parts that would be impossible or wildly expensive using conventional tooling.

From architectural panels and stone inlays to precision aerospace parts and delicate electronics, the waterjet's combination of strength and delicacy opens creative and practical doors.

2.2 CNC Waterjet: Not First, Never Last

If you're thinking about diving into CNC, a waterjet isn't usually the first machine people start with because of the high cost, maintenance, and sheer size of the equipment. But if you're running a shop or business that deals with multi-material projects, odd shapes, or requires cutting without heat distortion, adding a CNC waterjet can elevate your capabilities to a whole new level.

It's the kind of machine that doesn't just add to your workflow, it transforms what you can offer your existing and future clients. In short, CNC waterjet cutting is one of the most versatile, precise, and material-friendly processes in modern fabrication.

It's the secret weapon behind everything from custom car parts and luxury stonework to aerospace components, or multi-tools like my Drais BMTs (case study in the first Mastering CNC & Digital Fabrication).

If you want a machine that can quite literally cut (almost) anything, a CNC waterjet is the heavyweight champ of the shop floor. It has no flames or sparks, just raw, water-powered yet highly controlled precision.

2.3 Waterjet Safety Myths & Mistakes

The mighty CNC waterjet, the industrial machine that lets you slice through pretty much anything short of diamonds, slowly using nothing but water, garnet, and sheer pressure.

Here are some things to watch for.

Myth 1: "It's only water, I don't need PPE."

Right. Just like a fire hose is "only water," and we all know how gentle that feels when it slams you into a brick wall.

A waterjet operates at pressures up to 60,000 psi (≈ 4 100 bar), that's enough to cut through 6 inch (150 mm) steel using garnet of course. No matter how thick you think your skin is, it's going through it in very short order.

So no, you shouldn't stroll up in your flip-flops and sunglasses to check the nozzle. As with just about every other CNC platform, safety goggles, steel-toe boots, and appropriate shop wear aren't fashion accessories but a requirement.

Myth 2: "It's fully enclosed, so I don't have to pay attention while it cuts."

Look, enclosure or not, you are still operating a machine with enormous hydraulic forces, high-pressure lines, and ultra-fine abrasive.

Things can, and will go wrong: hoses can rupture, garnet can clog, nozzles can misalign, or parts can shift mid-cut with no warning.

Operating CNC equipment is like riding a motorcycle, it's not a matter of if an accident will happen, but when. The key is ensuring you have the right protection in place when it does. Leaving it running while you hit the coffee machine isn't "efficiency," it's gambling with very expensive chips.

Stay alert and monitor the job as you would with any other piece of equipment in the shop.

Myth 3: "A leak is just a leak; I'll fix it later."

A CNC waterjet leak is not like a dripping sink faucet. At extreme pressures; even a pinhole leak can turn into a mini water cannon, slicing through skin, flesh, and bone faster than you can say, "Wait, what was that hiss?"

Never, ever ignore a leak. Shut the system down, depressurize, and investigate immediately. Remember: high-pressure water doesn't ask questions, it just goes.

Myth 4: "You can just stick your hand in the tank, the jet's on the other side."

First, the jet isn't always where you think it is, and second, the aftermath of sticking your hand in a tank with an active cutter can land you on a very special episode of "World's Dumbest Injuries." That vat is full of such toxic sludge, it makes the Ganges River look like a mountain spring.

Even if the head is parked or raised, keep your body clear of the cutting area at all times. Those nozzles aren't smart, they won't stop for an oops.

Myth 5: "Hearing protection? It's just water splashing around."

A CNC waterjet is loud. Not "vacuum cleaner" loud, but we're talking jet engine, standing-in-an-industrial-barn-during-a-storm loud. Between the pump, the pressurized stream, and the abrasive cutting, you can easily hit noise levels above 90–100 dB. That's permanent-hearing-damage territory. Same as a running a router with a worn-out spindle!

Want to keep your hearing sharp enough to appreciate your finished parts? Slap on those earmuffs or plugs every time you fire up the beast.

Myth 6: "I can run whatever material I want."

This is a tricky one because yes, technically waterjet cutting is a cold process without heat or sparks unless you are cutting titanium.

But that doesn't mean cutting everything is safe. Some materials, like certain composites, plastics, or lead-based metals, can break down under the force of the jet, releasing nasty particulates or micro-fragments into the water or air.

Always check your material's compatibility before you turn your waterjet into a DIY hazardous waste generator.

Myth 7: "I don't need to worry about garnet dust, it's wet, so it's safe."

Garnet abrasive is a critical part of the cutting process, but when dry, it can become a fine, inhalable dust.

Sure, it's damp when you pull it from the tank, but once it dries out on the shop floor or during cleaning, it's ready to float up your nose and into your lungs.

Wear a dust mask when handling dry spent abrasive, and sweep or vacuum carefully. Nobody wants "industrial sand lung" on their resume.

CNC waterjets produce clean, precise cuts in stainless steel.

Myth 8: "The software has collision detection, no chance the head will crash."

Software is smart, but it's not psychic.

If you've input the wrong standoff height, failed to calibrate your Z-axis, or misaligned your workpiece, the head can (and will) slam into material, fixtures, or even the tank itself. All CNCs are dumb in this way.

And at waterjet speeds, that's not a gentle nudge. It's a metal-on-metal punch that can shatter nozzles, break assemblies, and cost thousands in repairs. Always check your setup, do a dry run if needed, and don't assume the software will catch every human error.

Myth 9: "Slurry disposal? It's just mud, I'll hose it down the drain."

Waterjet waste isn't just muddy water, it's a slurry of fine metal particles, garnet grit, and possibly hazardous material residues. That vat is a unique chemistry experiment each and every time it fills up with debris.

Dumping it down the drain can clog your pipes, wreck your local water system, and trigger fines or shutdowns faster than you can say, "I thought it was okay." Invest in proper filtration or disposal methods, and keep your shop clean.

Myth 10: "Maintenance? I'll do it after this big rush job."

I get it. You're busy, you've got deadlines, and the waterjet seems to be running fine. But ignoring routine maintenance on a machine that operates at extreme pressures is like ignoring the oil light on a race car.

Pumps wear, seals crack, filters clog, and hoses age, and when they fail, they fail big.

Don't turn your productivity hero into a hydraulic nightmare. Stick to the maintenance schedule like your business (and fingers) depend on it, because they do.

Waterjets sand through material instead of slicing it.

2.4 Lessons I've Learned About Waterjets

A CNC waterjet is an incredible tool: precise, powerful, and capable of cutting through materials that would destroy a router, frustrate a laser, or confound a plasma cutter. From titanium to stone to thick composites, it slices with a force that demands respect. But with great power comes great responsibility, and a high-pressure water stream doesn't care if you're a rookie or a pro; it'll bite anyone who's careless or overconfident around it.

So gear up properly, stay alert, and follow safety protocols to the letter. Respect the process, double-check your setup, and never take shortcuts. Laugh with the machine, not at it, because overconfidence is the quickest path to injury.

The best waterjet operators aren't the ones who cut the fastest or flashiest, they're the ones who finish every shift with all their fingers, hearing, and hard-earned reputation still intact. Most shop injuries are cumulative, not instant.

Now, go forth and cut. Smartly, safely, and with the pride of someone who knows what true craftsmanship looks like.

2.5 What's Wrong With My Waterjet?

Waterjets are majestic beasts, slicing through titanium with cold water (yes, it heats up from abrasion) and 60,000 psi of attitude.

But when your once-clean cut starts looking more like a seismograph, the machine isn't broken, it's just trying to tell you something.

"When the cut's ugly and the pressure's fine," it usually means your orifice and mixing tube are misaligned, worn out, or clogged with garnet confetti.

Your Waterjet Has Two Jobs: Focus and Blast

The waterjet is, at its core, a beautifully orchestrated one-two punch.

First, the orifice forms a razor-thin, supersonic stream of high-pressure water, a beam so powerful it can slice through metal like it's paper.

Then comes the mixing tube, also called the nozzle, which draws in abrasive particles (usually garnet) and focuses the jet into a concentrated, cutting force capable of carving through materials that would stop most other tools in their tracks.

But this precision dance depends entirely on the condition and alignment of those two components. If the orifice is worn out, the stream loses coherence.

If the mixing tube is misaligned, or worse, installed backwards (yes, that happens more often than you'd think), it can cause the abrasive flow to become chaotic, and the waterjet's power is scattered. What was once a scalpel becomes a soggy mess with bad aim to boot.

Instead of a clean, surgical cut, you get a jagged, frustrated path that looks like it was made by a machine having a very bad day. Of course, this will be the operator's fault to boot.

Warning Signs You're Not Cutting, You're Wandering

Wavy, tapered, or just plain ugly cuts used to slice clean like sushi now resemble uneven lasagna. If the bottom edge is tapering or the top edge looks sandblasted, your mixing tube is probably toast.

Inconsistent cut quality halfway through a job is a classic red flag. The first piece is flawless, the tenth is garbage. You didn't change anything, except maybe the pressure, the garnet, and your will to keep going. When your waterjet's mood shifts, blame the nozzle, not the stars.

Abrasive blowback shows up as dust in the tank or sudden jet sputtering. That usually means a blocked or misaligned orifice, or maybe you blew a seal (the mechanical kind, not a marine animal).

Unusual sounds are another warning sign. A healthy waterjet hums with pressurized precision. A failing orifice screeches like a dentist drill caught in a wind tunnel. If the machine starts making weird noises, it's not gaining sentience, it's begging for maintenance.

Know Your Wear Patterns (a.k.a. Cut Autopsy)

Every part of a waterjet plays a role in delivering clean, precise cuts, but like all hardworking components, they wear down over time. Recognizing how each part functions and how it typically fails can save you hours of troubleshooting and costly downtime.

Part	What It's For	How It Dies
Orifice	*Creates precise, narrow water beam.*	*Chips, cracks, becomes oval, scattered jet.*
Mixing Tube	*Focuses abrasive stream.*	*Inner diameter erodes, sloppy, weak jet.*
Abrasive Line	*Delivers garnet.*	*Kinks, clogs, or launches like an unhinged hose.*
Focus Assembly	*Holds it all together.*	*Misaligned, comes loose, suffers heat damage.*

How to Inspect Without Crying

The Orifice

Whether it's ruby, sapphire, or diamond: should look like a perfectly centered dot of waterjet perfection. Peek through and see anything but a clean circle (chips, cracks, an oval)? It's not mysterious. It's done.

Even diamond orifices, the overachievers of the group, wear out eventually. Precision loves alignment, and so does your orifice.

The Mixing Tube

The mixing tube is supposed to have a tight, focused inner diameter, usually around 0.030 - 0.050 inches (≈ 0.76 to 1.27 mm). If yours has widened to double that, congratulations: you're no longer cutting steel, you're power-washing oatmeal.

If you can't see the jet from one side, it's not stealth mode, it's just worn out. When your kerf is wider than your toolpath, that tube isn't being subtle. It's screaming for retirement, same idea as a cracked CNC laser lens.

Practical Tips to Stay on the Beam

Don't overtighten the orifice, it's a precision gemstone, not a lug nut. Treat it with care. Alignment matters, too. If it's even slightly off, the waterjet slams into the mixing tube wall and sprays out like a busted pressure washer at a car wash from hell.

Flush your abrasive lines daily. Skipping that means clumped garnet, clogged nozzles, and a machine that starts making noises like it's plotting revenge. And don't cheap out on garnet. Low-quality abrasive shreds performance and chews through mixing tubes faster than you can say "unexpected downtime."

If your mixing tube wore out early, it didn't just fail. Let's be honest, you murdered it with extreme prejudice.

How Long Should Parts Last?

Waterjet parts don't last forever, and pretending they do is a great way to wreck a job mid-cut. Each component has a shelf life, whether it's a delicate orifice or a workhorse mixing tube. Knowing when to expect failure (and what really shortens that timeline) can mean the difference between smooth cutting and an unexpected shutdown.

Part	Lifespan	Reality Check
Ruby Orifice	*40–100 hours*	*Cheap but fragile. Treat it like a crystal, not a tool.*
Diamond Orifice	*500–1,000+ hours*	*Pricey but stable, just don't misalign it.*
Mixing Tube	*60–120 hours*	*Lifespan swings with abrasive quality and pressure.*
Garnet Filter	*Change monthly (or more)*	*Unless you're into surprise clogs and upstream chaos.*

Shop Habits That Keep You Cutting Clean

Like CNC plasma or welding contact tips, label old and new orifices clearly so worn-out ones don't sneak back in like waterjet zombies. Keep spare parts on hand, these machines love to break when it matters most. Store orifices in a dry box and clean your abrasive hoppers and lines regularly, clumps kill consistency.

Focus the Jet, or Face the Dross

Waterjet systems are miraculous, until they aren't. Most bad cuts don't need a system overhaul or deep programming dive. They usually trace back to a worn or misaligned orifice, eroded mixing tube, tired swirl path, or a clogged abrasive line staging sabotage. Before tweaking feedrates or rewriting G-code like a mad scientist, pause and squint at the nozzle.

Nine times out of ten, when a waterjet acts up, the problem isn't upstream, it's in your hand.

❏ Let the Garnet Flow, It's Q&A Time!

2.6 How does abrasive grain size effect cut quality?

When it comes to waterjet cutting, abrasive grain size is a big deal. It's like choosing between a scalpel and a sledgehammer, both get the job done, but the aftermath's a little different. Fine grains, like 120 mesh garnet, are the perfectionists of the bunch. They deliver crisp, clean edges and love delicate jobs such as glass, ceramics, tiny parts, basically anything you'd rather not shatter. But don't expect them to be quick about it. They cut slow, generate extra heat, and wear out your nozzles like they're on a mission to sabotage your maintenance budget.

Coarse grains, on the other hand, such as 60 or 80 mesh, are the brutes. They blast through thick steel and tough aluminum like they've got somewhere to be. They're fast, furious, and not exactly subtle, leaving behind a rougher finish that might need a cleanup crew (or at least some post-processing).

Most shops play it safe with 80 mesh, the middle child of abrasives. Not too fast, not too pretty, but reliable. Just remember: grit quality matters too. Crummy, inconsistent abrasive is like putting gravel in your espresso machine, expect clogs and chaos.

At the end of the day, think of "mesh" as grit for sandpaper, the lower the number, the harsher, but faster the cut.

2.7 What are the common pump maintenance tasks?

Waterjet pumps operate at extreme pressures, often exceeding 60,000 psi (≈ 4 100 bar), making routine maintenance critical for performance, safety, and cost management.

Daily maintenance tasks include inspecting for visible leaks, checking fluid levels, and listening for unusual sounds like squealing or knocking, which can signal bearing or seal wear.

Your waterjet pump isn't a magical fountain of infinite pressure, it's a hardworking beast that needs a little love to keep from throwing a hydraulic explosion. That's why smart operators keep one eye on the pressure gauges. If it's out of range, something's wrong, and your machine will let you know... usually at the worst possible time.

Every week, show it some TLC: clean or swap those high-pressure filters, give seals and check valves a good look, and make sure the hydraulic system isn't trying to quietly leak its way into early retirement.

Got an intensifier pump? Congrats, now you also get to baby the rods, seals, and check valves. Ignore them and you could be starring in your own personal disaster movie.

Monthly or quarterly (depending on how wild your cutting schedule is), it's time for deeper care: replace worn seals, crack open that intensifier and make sure its insides aren't turning to mush, check accumulator pressure, and make sure your cooling system is still actually cooling.

Some shops even run pump performance tests, like a stress test, but for your wallet if it fails.

Your part takes a shower but it's easily cleaned.

2.8 How do you calculate kerf compensation for intricate designs?

Kerf compensation sounds fancy, but it's really just accounting for the width your waterjet chews out of the material, aka, the kerf. With waterjets, that's usually somewhere between 0.030 - 0.050 inches (≈ 0.76 - 1.27 mm), depending on factors like your nozzle diameter, abrasive grit, water pressure, and the thickness or type of the material you're cutting. Ignore kerf, and you'll end up with holes too small, parts too big, and fits that only work if you're forcing things together with a mallet (not recommended unless you enjoy making your life difficult).

To dial in kerf compensation, smart operators start with a test cut, not because they like wasting material, but because precision matters. This process should be familiar with anybody running a CNC router.

Measure that real-world kerf width, feed it into your CAM software, and voilà, the software offsets your toolpath by half the kerf on each side. That way, what you cut actually matches your drawing, not some mystery blob vaguely shaped like your part.

And if you're working on intricate or tight-tolerance parts, even tiny kerf miscalculations can mess everything up. A small error on a complex internal shape can ruin the fit, throw off your whole assembly, or just make your work look like you eyeballed it with a butter knife.

Bonus round: some modern CAM programs can apply dynamic kerf compensation, adjusting the offset on the fly based on material thickness, corner radii, or cut direction. This is especially useful when switching between inside and outside profiles, so your internal holes don't end up way too snug, and your outer edges don't look like they got into a bar fight with the nozzle.

Long story short? Proper kerf compensation is what separates a clean, accurate part from one that ends up as a very expensive learning experience. Kerf is inevitable, but getting it right is totally within your power.

2.9 How does pierce time affect brittle materials?

Pierce time is how long your waterjet focuses in one spot to punch through the material before following the cut path.

Sounds simple, unless you're cutting something brittle like glass, ceramics, or carbon fiber. Then it's more like defusing a bomb. Laminated material just up the challenges.

If pierce time is too short or pressure's too high, the jet hits like a tiny wrecking ball, sending shockwaves that brittle materials can't handle. Cue cracking, shattering, or highly possible delamination.

The fix? Use "low-pressure pierce" modes that ease the jet in gently. Some advanced machines even ramp up pressure gradually, and I'm sure as technology develops, something "newest and greatest" will also be coming out solving problems to issues we don't even know existed.

For extra-sensitive materials, pre-drill starter holes so the jet doesn't have to muscle through. Support backers like foam, MDF or rubber also help absorb impact and reduce vibrations.

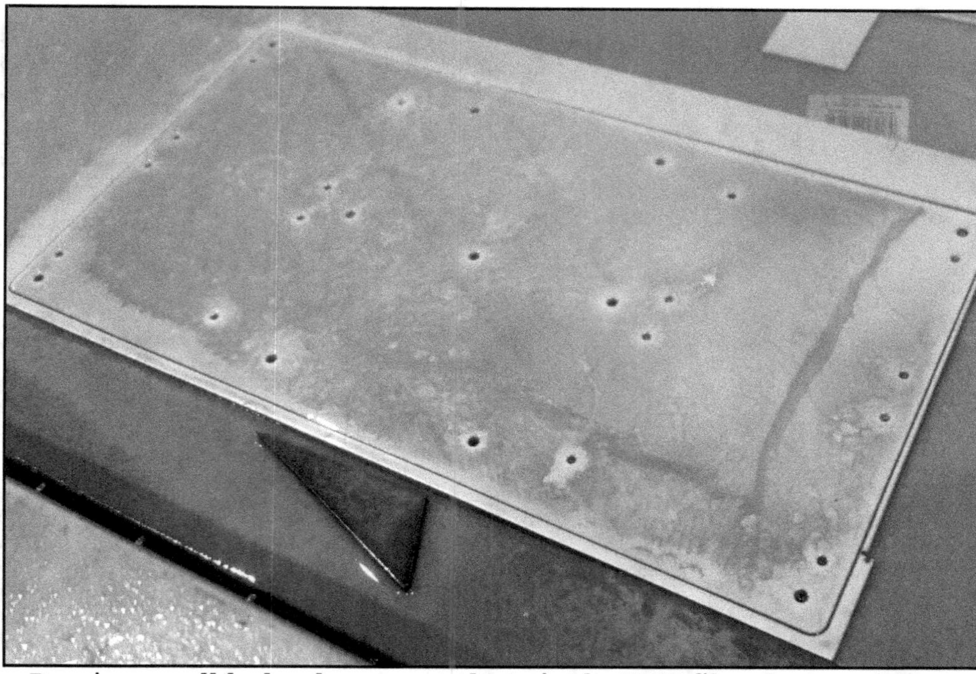

Precise small hole placement that rivals CNC fiber laser cutting.

2.10 What nozzle alignment methods improve precision?

In waterjet cutting, nozzle alignment isn't optional, it's ***EVERYTHING***. If the jet isn't hitting the material square-on, you'll get tapered cuts, sloppy edges, inaccurate dimensions, and worn-out parts without rhyme or reason.

To keep things on point, operators use tools like laser alignment systems (same used for all high quality industrial CNCs), dial indicators to check flatness, and jigs or calibration blocks to square things up. It's like giving your nozzle a chiropractic adjustment, for accuracy.

Alignment checks are a must after maintenance, collisions, or swapping of parts. Even tiny misalignments can snowball into big errors over long or intricate cuts. Plus, a properly aligned nozzle lasts longer and cuts cleaner.

Like anything else in custom fabrication, you always want to measure twice and cut once... but align first, or risk turning your precision machine into a major headache.

CNCs are like cats: they'll warn you before they destroy something, but you probably won't listen in time.

Waterjets are well suited for producing fine, small stencil features.

2.11 How do waterjet pressures compare for thin vs. thick materials?

Waterjet pressure is a big deal as it's the muscle behind every clean cut. But more pressure doesn't always mean better. For the relatively thin stuff, blasting at 60,000 psi (≈ 4 100 bar) is like using a fire hose to water a houseplant.

Running at 30,000–40,000 psi (≈ 2 100–2 800 bar) usually gets the job done without launching your parts into outer space or leaving shredded edges behind.

Thicker, tougher materials? Now's the time to unleash the beast. Cutting 2 inch (≈ 50 mm) steel or chunky aluminum needs serious muscle, up to 60,000 psi (≈ 4 100 bar) or more, to push through cleanly. Skimp on pressure, and you'll get sluggish cuts, tapered edges, striations, and maybe a part that's technically still attached.

But pressure isn't the whole story. Nozzle size, abrasive type, feed rate, and standoff distance all work together like a high-stakes pit crew.

Nail the setup, and you'll get crisp edges and fast performance, like anything else, practice makes perfect.

2.12 What are the challenges of cutting laminated materials?

Cutting laminated materials with a waterjet is like trying to slice a fancy seven-layer cake with a firehose, sure, you can, but things can go sideways fast. Whether it's layered composites, laminated glass, or those mysterious multi-material panels that look like they were designed by a bored engineer with glue, they all come with one major issue: delamination.

That high-pressure jet, packed with abrasive grit, doesn't just cut, it likes to sneak between layers, lifting, fraying or bubbling. The pierce phase? That's where chaos begins. Hit it too hard, and your nice layered sheet can peel apart like a soggy sandwich. That's why pros ease in with low-pressure pierces or pre-drill starter holes for less initial shock.

Feed rate matters too: go too fast and you'll rip the layers like a toddler with wrapping paper; go too slow and the material turns into a sponge, soaking up water like it's trying to ruin your weekend.

Want to play nice with laminates? Use a support backer (foam or sacrificial MDF board), pick a finer abrasive to avoid violent grit-flinging, and as always, test first.

Laminates with adhesives or interlayers have minds of their own and like to surprise you mid-cut.

2.13 How do you manage slurry disposal responsibly?

Waterjet cutting is awesome, but it leaves behind a lovely mix of water, fine material particles, and used-up abrasive (usually garnet). It's a gritty, gooey mess known as slurry.

While it might look like industrial soup, you definitely don't want it going down the drain. Your plumbing, your wallet, and the environment will all scream in protest.

Step one: separate the mess. Most shops use settling tanks or filtration systems where the heavy stuff, such as garnet, metal bits, and mystery sludge, sinks to the bottom.

The (somewhat) cleaner water on top can be reused or discharged if it meets local rules. Some setups even have filter presses or centrifuges that compress the gunk into waste "cakes" that are way less fun than they sound.

Want to be fancy (and save money)? Some shops install garnet recycling systems. These clean and reuse a portion of the abrasive, but don't get too excited. Garnet breaks down over time and eventually becomes too wimpy to cut anything but your hopes and dreams.

Think of the "sand" in your sandblaster, it eventually becomes useless powder where all the grit is gone.

Whatever you do, do not dump slurry into a drain as it's the best way to clog a drain. "Sand" doesn't exactly flow well in a constrained environment.

2.14 How does taper compensation improve accuracy?

Taper in waterjet cutting is the slight angle that happens because the jet loses energy as it cuts downward, meaning the top of the cut is a bit wider than the bottom. No big deal for rough work, but for precision parts, it's the reason things don't fit and tolerances get out of whack.

Enter taper compensation, the machine's way of fighting back using the power of software. With tilt-heads or smart software, the nozzle angles just enough during the cut to keep those edges near-vertical. It's especially crucial for thick metals, or tight-tolerance parts.

Without compensation, you're stuck oversizing holes, tweaking CAD files, or breaking out the end mill for a second round. All of which cost time and money.

Thankfully, most modern systems offer dynamic taper control, adjusting on the fly for different materials or thicknesses. This "idea" is also shared with other CNC kin.

It might add a bit of complexity, but taper compensation means cleaner fits, fewer do-overs, and no surprise wedges when your "perfect" part doesn't quite line up.

2.15 How does garnet recycling affect the cutting?

Recycling garnet can save you money, but it comes with a performance hangover. Once-angular grains turn into smooth pebbles after a cut or two, which means slower cuts, rougher edges, and striation lines that scream "budget job."

To keep quality from tanking, many shops mix recycled and fresh garnet 50/50. A compromise between cutting power and cost savings. Going full recycled only works for rough cuts or parts where no one cares what the edges look like.

There's also the issue of junk, recycled garnet can collect fine debris and mystery bits that clog nozzles or wear down your mixing tube. Good filtration is a must unless you like surprise maintenance.

2.16 What environmental factors affect waterjet cutting efficiency?

Waterjet cutting isn't just about the machine, the shop environment can make or break your results. Cold temps thicken hydraulic fluid, slowing everything down and hogging energy. Hot temps overheat pumps, fry seals, and may trigger thermal shutdowns.

Humidity? Also a troublemaker. It can cause condensation in electronics, rust in the works, and soggy garnet that clumps up and jams feed lines like a bad breakfast.

Water quality matters too. Hard water leaves mineral gunk in your pump and nozzles, wrecking seals and killing flow. Use softened or filtered water unless you enjoy surprise maintenance bills.

And don't ignore shop cleanliness as airborne dust can sneak into tanks and hoppers, leading to contamination and lousy cuts. Keep the shop clean, cool, and dry, and your waterjet will return the favor with faster cuts and fewer breakdowns.

At *CNCROi.com*, I run multiple CNC machine platforms basically next to one another, not ideal but you make due with what you have when you have it.

2.17 How do you reduce edge chipping cutting glass?

Glass loves drama, cut it wrong, and it chips, cracks, or just explodes. Even though waterjets cut cold, the abrasive jet still packs a punch.

First tip: go fine with 120 mesh garnet. Coarse grit cuts faster but chews up edges like sandpaper. Next, ease into the pierce with low pressure or a gradual ramp-up keeps the jet from shocking the glass into a thousand pieces. Bonus move: pre-drill a hole and skip the pierce altogether.

Speed matters; too fast, and you chip; too slow, and the jet digs in and wrecks the edge. Support the glass with rubber or foam backers to keep vibrations under control.

2B finish 304 stainless in 12 gauge makes very durable stencils.

Cut. Rinse. Ship.

2.18 How can 5-axis waterjets handle beveled cuts?

A 5-axis waterjet doesn't just cut straight down, it tilts, swivels, and struts its stuff like a cutting head doing yoga. This lets it tackle bevels, countersinks, and funky geometries that would otherwise be impossible.

Want angled edges or precision tapers? The 5-axis head tilts mid-cut, following fancy CAM instructions. It's especially handy in fields like aerospace, where parts need to fit together exactly, not "close enough if you squint."

Of course, this level of fancy comes with complexity. Programming beveled cuts isn't plug-and-play, you'll need advanced CAM software and a solid grip on calibration. Plus, all that tilting and rotating adds mechanical stress, so the machine has to be built like a tank. Nozzle wear also ramps up since the jet's hitting the material at all kinds of weird angles that may spray back into strange spots.

2.19 What are the costs of abrasive per cut inch?

Abrasive isn't just sand, it's money flying out of your nozzle. The cost per cut inch depends on what you're cutting, how thick it is, how fast you're going, how nice your garnet is, and how efficient your setup is. Costs vary greatly on the quality, quantity and geometry of the garnet. Thin plastic? Cheap. Thick heat treated 62 HRC tool steel plate? Cry.

To get nerdy with it: you need to know your abrasive flow rate (in pounds per minute), your cutting speed (in inches per minute), and what you're paying per pound of garnet. Say you're using 1 lb/min and cutting at 10 in/min, that's 0.1 lbs/inch. At $0.50/lb, you're spending $0.05 for every inch of cut glory. This might represent a lot or nothing at all, it all depends on your profit margins and the machine's capabilities regarding projects you take-on.

Want to cut costs without cutting corners? Dial in your flow rate, consider a garnet recycling setup, or find that magical balance between cutting fast (more abrasive, less machine time) and cutting slow (less abrasive, but more labor and machine wear).

2.20 How do pump technologies (direct drive vs. intensifier) compare?

In the world of waterjet cutting, your pump is the powertrain, and there are two main types in the ring: Direct Drive and Intensifier. Both bring muscle, but in very different ways. Are you a bodybuilder or a powerlifter?

Direct Drive pumps are the minimalist athletes like bodybuilders. They use crankshafts and plungers driven by electric motors to deliver high-pressure water with fewer moving parts and solid energy efficiency.

They're simpler, cheaper to maintain, and perfect for light-to-mid workloads. Think thin metals, plastics, and hobby-shop heroes. But they usually cap out around 55,000 psi (≈ 3 800 bar), and under constant industrial use, components can wear out faster.

Intensifier pumps, on the other hand, are the hydraulic powerlifters. They amplify pressure to 60,000 psi (≈ 4 100+ bar) using fluid-driven pistons, making them ideal for thick metals, multi-head setups, and round-the-clock jobs.

They deliver consistent pressure even when cutting loads change and often come with flow-smoothing perks like pressure dampeners.

The trade-off? Complexity. More parts, more seals to baby, and a higher up-front cost. They're the high-maintenance sports car of waterjetting, powerful, but needy.

Put another way:

- Heavy-duty shops = intensifier
- Smaller, budget-conscious shops = direct drive

Choose the one that matches your workload, and your tolerance for overpriced maintenance calls.

As a side note, make sure to ask your sales rep if you can attend a pump rebuild, this will happen, might as well get an idea of what's involved and more importantly, the costs and downtime they will cause.

2.21 What is the role of catch tanks in waterjet setups?

Catch tanks sit under your waterjet like a giant shock absorber, soaking up the leftover fury from the jet once it blasts through your material. Without one, you'd be turning your machine base, or worse, your shop floor, into a hole via high-pressure destruction over time.

Filled with water, the tank acts like a hydraulic sponge, calming the chaos. Abrasive and debris settle in the bottom, so yes, it needs cleaning before it turns into a gritty swamp. Some systems even have grates to catch the big chunks before they sink, again, ask your sales rep for options.

Skip maintenance, and trouble brews: reduced water depth, overflow, pump pressure issues, or even the nightmare scenario, jet splashback that cooks your cutting head.

Bonus: catch tanks also muffle the waterjet's angry roar. Keep it clean and happy, and it'll quietly keep your cuts sharp, your machine alive, and your floor intact.

Marble is no match for CNC waterjet cutting.

2.22 How do you prevent delamination when cutting composites?

Cutting composites with a waterjet isn't a walk in the park. Mess it up, and you get delamination, where layers peel apart like a bad sunburn in July. Carbon fiber, Kevlar, honeycomb panels, they're all prone to splitting under pressure (literally).

To avoid this, pierce gently (I realize this is a repetitive theme with waterjets). Full pressure straight out of the gate can force water and grit between layers, breaking bonds and fraying edges. Use low-pressure pierce modes, starter holes, or pressure ramp-up tricks to sneak in without scaring the material.

Speed matters too; too fast and you tear the top layer; too slow and water soaks in like a sponge. Sharp, fine abrasive helps reduce tearing, and a foam or rubber backer keeps vibrations from turning your panel into a trampoline.

Some composites even need multiple passes at lower pressure to stay intact. And test cuts? Non-negotiable. What works on one mystery material may fail spectacularly on the next. I hope you start to realize how similar things between different CNC platforms are.

2.23 What safety measures are critical when operating at 60,000 psi (≈ 4 100 bar)?

Operating a waterjet over 60,000 psi (≈ 4 100 bar)? That's steel-cutting, limb-endangering power, so safety isn't optional. Start with PPE: safety glasses, steel-toed boots, ear protection, and sometimes a face shield if you are working close.

Before any maintenance, use proper lockout/tagout (LOTO). Never touch pressurized lines, a pinhole leak can inject water like a high-speed syringe. Not fun.

Inspect hoses, seals, and fittings regularly. If it's cracked, bulging, or held together with hope, replace it. No "temporary fixes" and operators need real training too.

2.24 How do CAD/CAM settings impact waterjet cycle times?

When it comes to waterjet cutting, your CAD and CAM settings are like the difference between a smooth highway drive and a GPS meltdown in the woods. Get them right, and your machine hums. Get them wrong, and you're watching it zigzag like it's lost.

Toolpaths are everything. Smart paths mean fewer useless moves, less jet wandering around like it's sightseeing, and more actual cutting. Minimize air cuts and smooth out those transitions because time is money, and your jet's not paid by the hour.

Lead-ins and lead-outs? Think of them as graceful entrances and exits. Set them right and your part looks sharp. Set them wrong and suddenly you've got weird scars and wasted time. It's not art. It's just bad programming.

Then there's nesting, the Tetris of fabrication. Pack those parts tight, and your machine travels less, cuts more, and you waste less material. Bonus: fewer pierces, faster jobs.

Pierce settings also matter. Ramp the pressure smartly, don't rush brittle materials, and avoid turning every pierce into a dramatic event. For jobs with a million little parts, efficient piercing is your best friend.

Modern CAM tools even let you simulate everything first, regardless of the CNC platform, so you can quickly spot inefficiencies before the jet (or laser, or router, or EDM, etc.) starts dancing off-script.

2.25 What material combinations are unsuitable for waterjet cutting?

Waterjets, like any other tool in a custom fabrication shop is amazing, but they're not magic.

While they can slice through a wild variety of stuff, some materials just don't play nice. Let's talk about the ones that can throw you off.

Tempered glass? Instant chaos. It's built to shatter under stress, and a 60,000 psi (≈ 4 100 bar) waterjet is pure stress. Result? Boom. It explodes like glitter at a toddler's birthday. Clean cuts? Nope.

Laminated or layered materials, like composites or bonded metals, can also misbehave. If the glue's weak, the jet sneaks in and peels them apart like a soggy sandwich.

Porous stuff like foam or open-weave fabric? It soaks up water like a beach towel, turning clean cuts into wet disasters. And some plastics (especially low-melting ones) can warp or sag under longer cuts. Sure, waterjets are "cold," but not cold enough for these drama queens.

And don't forget reactive metals like magnesium alloys, they can spark or combust. Rare, but not the kind of surprise you want during a shift.

Know your material. Run test cuts. Just because something looks cuttable doesn't mean it won't fight back.

Use plywood or economical MDF as reliable waterjet backing.

2.26 How does abrasive flow rate calibration impact cutting efficiency?

Calibrating your abrasive flow rate is key to waterjet efficiency. Too low, and the jet just kind of tickles the material: slow cuts, rough edges, and striations galore. Too high, and you're blasting through garnet (a pricey consumable) like confetti at a parade, with little to no gain in speed or quality.

The sweet spot depends on your material, thickness, nozzle size, and edge finish goals. Thick steel? Sure, crank up the flow. Thin plastic or aluminum? Much less will do, no need to overfeed the hourly consumables beast.

Operators usually dial it in with test cuts or by checking the manual (yes, the one no one reads). Also, don't forget to keep an eye on the garnet system as clogs, moisture, or uneven feeds can throw everything off and leave you wondering why your cut looks like it was done with a rusty spoon.

Nail the flow rate, and you'll get clean cuts, save money, and keep your machine running smooth. Miss it, and you're either wasting garnet or cutting like you're out of juice.

2.27 What factors determine optimal standoff distance?

Standoff distance, the gap between your nozzle and the material, might seem small, but get it wrong and things get ugly fast. Same holds true with other CNCs and especially welding processes!

Too low, and you risk nozzle crashes, fixture collisions, or splashback clogging the mixing tube. Too high, and your jet loses focus like a laser pointer on a disco ball, weaker cuts, more taper, and jagged edges.

The sweet spot? It really depends on your nozzle, material, and thickness. Harder or thicker stuff such as tool steel likes a tight gap to keep jet energy sharp. Softer or delicate materials like foams may need a little breathing room to avoid damage or distortion.

Worn nozzles or mixing tubes can also mess with jet behavior, so standoff might need tweaking. And if you've got a dynamic cutting head doing tilt tricks, factor that in too as angles can change everything.

Smart operators use gauges or spacers to dial in standoff consistently. Nail it, and you'll get clean cuts, longer nozzle life, and results that don't make you say, "What the heck happened here?" Hint, it's you, not the machine.

2.28 How can you handle micro-cutting on delicate materials?

Micro-cutting is where your waterjet ditches brute force and puts on surgical gloves. Creating ultra-fine features in delicate materials takes finesse, patience, and settings that don't scream "maximum power!"

First off: lower the pressure. While standard jobs run at 50,000–60,000 psi (≈ 3 400–4 100 bar), micro-cutting often drops to around 30,000 psi (≈ 2 070 bar) or even less. Thin metals, fragile composites, and materials with anxiety issues just can't handle full-blast erosion.

Use finer abrasive, like 120 mesh. Coarse grit is like chucking gravel at a soap bubble, not ideal for detail work. Finer grit gives you cleaner edges without the carnage. Slow down the feed rate. The jet needs time to gently carve through tight curves and delicate geometry. Going too fast is like trying to ice a cake with a power washer.

Pierce carefully. Use low-pressure pierce modes or pre-drilled starter holes, blasting full pressure right away is a great way to blow out tiny parts before the cut even begins.

Support the material properly. Any flex, jiggle, or wobble can turn crisp micro-details into a sad, blurry mess. Use foam, clamps, or even sacrificial backing to keep it rock solid. If you're doing micro-cuts regularly, upgrade to specialized nozzles and fine-tuned CAM software.

These tools keep your speed in check, your accuracy sharp, and save you from diving headfirst into a sea of headaches.

2.29 What's the role of water hardness in pump longevity?

Water quality, especially hardness, has a bigger impact on your waterjet than you'd expect. Hard water's full of minerals like calcium and magnesium that love to settle inside high-pressure lines and pumps like squatters. Over time, they create scale buildup that increases friction, wrecks flow, and wears out seals, plungers, and orifices.

Smart shops install water softeners or filtration systems to keep that mineral menace out. Many manufacturers even specify strict limits for water quality, usually measured in parts per million (ppm) of total dissolved solids (TDS). Ignoring them can void your warranty or earn you some very expensive repair bills.

Regular maintenance, like checking for scale and flushing the system, is key. Skip it, and that shiny pump built to last thousands of hours might be limping along in half the time.

2.30 How do you prevent nozzle clogging?

Nothing ruins a long waterjet cut like a surprise nozzle clog. The usual villains? Moisture, dirty abrasive, and inconsistent flow. It's generally a combination of things working in perfect alignment when you have a project that's due yesterday that you are doing the night before.

To stop the madness, keep garnet in sealed, dry containers, and use dehumidifiers or desiccants in the feed line, soggy grit clumps like cereal in milk and jams everything up.

Regularly check and clean your abrasive delivery system, such as hoses, feeders, metering devices before buildup turns into a full-blown blockage. Use high-quality abrasive too; dusty, recycled grit is a clog just waiting to happen.

Watch for nozzle wear as worn orifices and tubes mess with flow and raise clog risk. Replacing them on schedule beats wrestling with mid-job downtime. For marathon cuts, some shops use smart sensors to detect flow drops and pause the job before chaos strikes.

2.31 What software features help simulate waterjet cut paths?

Modern waterjet CAM software is like having a virtual dry run, minus the mess. One of its biggest perks? Toolpath simulation. It shows exactly how the jet will move, where it'll pierce, and how lead-ins and lead-outs will play out, all before a single drop of high-pressure water flies.

This virtual preview helps you catch collisions, redundant moves, and wasted time. Additionally, some software even predicts taper, edge quality, and cut time. Fancy, right? You can even watch material get "removed" in a 3D simulation, ensuring no area gets skipped or undercut.

Need to save material? Automatic nesting arranges your parts like Tetris pros to reduce waste. And pierce sequencing makes sure tiny parts get cut last so they don't shift mid-job and ruin everything.

2.32 How does table size affect cutting tolerances?

At first glance, a bigger waterjet table just means you can cut larger sheets... yay, right? But here's the twist: more size = more room for error. Over long spans, tiny issues like mechanical deflection, thermal expansion, or vibration can snowball into noticeable accuracy problems. Yep, it's the same issue across all larger CNC platforms.

If the table's even slightly unlevel, the cutting head might drift over distance, introducing taper or positional quirks. That's why big tables need beefier gantries, precision rails, and regular calibration. Ideally with dial indicators, lasers, or good old-fashioned test cuts.

Also, material support becomes a bigger deal. A large sheet that isn't fully supported can sag or flex, making your cut about as straight as a banana. More supports, clamps, or vacuum fixtures can help keep things steady.

Something you may not have thought about is that a larger table also means you're accumulating a lot more debris in the catch tank, which will take longer to clean-out.

CNC waterjets are capable of very fine detail machining.

Explore unlimited creativity, even in two dimensional CNC design.

2.33 How can waterjet systems be adapted for thick composite panels?

Cutting thick composite panels, like multi-layer carbon fiber, Kevlar, or honeycomb sandwiches, is where waterjet cutting starts to feel more like composite surgery. The main enemy? Delamination, a.k.a. the layers saying "nope" and peeling apart under pressure.

To avoid this, as mentioned previously, drop the pierce pressure or pre-drill starter holes to avoid hitting the material like a freight train.

Finding the right cutting speed is critical. Move too quickly, and the jet doesn't fully penetrate the lower layers. Move too slowly, and the panel absorbs too much water, causing resins and adhesives to soften and the material to deform. Enter: the fine-grit abrasive. It's gentler, sharper, and less likely to turn edges into a fuzzy mess.

Some shops go with multiple passes, rather than forcing the cut in one go, they let the jet take smaller bites. It's slower, sure, but it's way easier on both the machine and the material (and your nerves). The same approach is used for other CNC platforms as well.

Now let's talk fixturing. Thick panels love to flex at the worst possible moment. If they shift mid-cut, your precision part suddenly looks like modern art.

Clamps, supports, and even sacrificial backers (like foam or MDF) help absorb backsplash and keep pressure evenly distributed. Heavy "blocks" of metal along the edges also helps keep things down and static.

Success here isn't about brute force, it's about finesse. Proper calibration, smart settings, and a deep understanding of the material (plus a few test cuts and maybe a little swearing) are what separate clean cuts from composite chaos.

Moral of the story? Treat thick composites with care because when carbon fiber fights back, nobody wins.

2.34 What are signs of excessive pump wear?

Your CNC waterjet pump is the system's heartbeat, like the tube is to a CNC laser or a spindle is to a CNC router, and ignoring its cries for help can get expensive fast. If pressure drops for no good reason, it could mean worn seals, tired rods, or lazy check valves phoning it in.

Slower cuts? That's your machine compensating for lost efficiency. Hydraulic fluid puddles, drips, or mysterious stains? Gasket or seal failures say hi.

Weird noises (grinding, squealing, knocking)? Could be bearings or plungers throwing a tantrum. Extra vibration or rising temps? That's the system yelling, "Please fix me."

Pro tip: Log pressures, fluids, and temps. Catching wear early keeps you cutting, not calling for parts and praying for rush next-day shipping.

2.35 How do you balance cut speed vs. abrasive consumption for cost savings?

Balancing cut speed and abrasive use is waterjet economics 101. Cutting faster eats more abrasive per minute (cha-ching!) but reduces machine time. Cutting slower saves on grit but racks up labor, wear, and project time.

The "right" choice depends on the job. High-precision work? Go slow with less abrasive for clean edges. Bulk parts or rough cuts? Crank up the speed and abrasive to boost throughput.

Smart shops don't guess, they run cost-per-inch math, factoring in abrasive price, machine wear, labor, and overhead. Advanced software helps too, adjusting speed and flow on the fly based on material and geometry.

The secret sauce? Test settings, track performance, and aim for that sweet spot where speed, quality, and cost all high-five. Running a waterjet (or any CNC for that matter) is not about being fast or cheap; it's about being smart.

🎖 2.36 CNC Waterjet: Top 10 Takeaways

Abrasive grain size shapes cut quality: *Fine grains give smooth, slow cuts; coarse grains cut faster but rougher. Choose based on material type and finish requirements.*

Pump maintenance matters: *At 60,000 psi (≈ 4 100+ bar), daily checks, clean filters, and seal inspections prevent expensive failures and keep the system running strong.*

Kerf compensation is critical: *The jet removes material as it cuts, so CAM software must offset the toolpath to ensure final part dimensions stay accurate.*

Pierce time affects brittle materials: *Fragile materials like glass need slow, low-pressure pierces or starter holes to avoid cracking or shattering.*

Nozzle alignment boosts precision: *A perpendicular nozzle ensures straight, accurate cuts. Misalignment leads to taper, bad edges, and wasted material.*

Pressure settings depend on thickness: *Thin materials need lower pressure to avoid blowout; thick ones require full force for clean cuts.*

Laminated materials are tricky: *Without fine-tuned settings, layers can delaminate. Slow pierces and backing support help keep them intact.*

Manage slurry disposal responsibly: *Spent garnet and water must be filtered and properly disposed of; dumping it can clog pipes and earn fines.*

Taper compensation improves accuracy: *Tilt-heads and smart software adjust for natural taper, keeping edges vertical and precise.*

Garnet recycling saves costs, with caution: *Reused abrasive cuts slower and leaves rougher edges; blending with fresh grit helps maintain quality and speed.*

✖ 2.37 Section Quiz

Treat this quiz like a waterjet, stay on track, or you're in for a high-pressure surprise!

1. What abrasive is most commonly used in waterjet cutting?
   A) Silica sand
   B) Diamond dust
   C) Aluminum oxide
   D) Garnet

2. What determines the kerf width in waterjet cutting?
   A) Abrasive type only
   B) Pump temperature
   C) Water hardness
   D) Nozzle diameter and standoff distance

3. Why is taper compensation important in waterjet cutting?
   A) To control heat-affected zones
   B) To balance abrasive flow
   C) To correct for angled cuts and achieve straight edges
   D) To reduce water usage

4. What happens if you cut too quickly on thick material?
   A) You save abrasive
   B) You reduce taper
   C) You leave incomplete cuts or rough edges
   D) You improve edge finish

5. What maintenance task is critical for pump longevity?
   A) Replacing the high-pressure seals
   B) Changing the abrasive type
   C) Cleaning the cutting table daily
   D) Polishing the nozzle

6. What is the purpose of the catch tank in a waterjet system?
   A) To hold abrasive before cutting
   B) To collect spent water and abrasive
   C) To store extra pressurized water
   D) To reduce taper

7. How does standoff distance affect cut quality?
   A) Too close breaks the nozzle, far blurs the cut
   B) Too far increases water pressure
   C) Too close improves pump lifespan
   D) It has no significant effect

8. What is one key environmental advantage of waterjets?
   A) No waste material produced
   B) No heat-affected zone
   C) Generates no noise
   D) Does not require electricity

9. Why is garnet recycling used in some waterjet shops?
   A) To improve cut speed
   B) To reduce abrasive costs and environmental waste
   C) To clean the cutting tank
   D) To increase nozzle life

10. What kind of cut edge is typical on the bottom of thick waterjet cuts?
   A) Smooth and polished
   B) Burned and melted
   C) Slightly rough with striations
   D) Brittle and cracked

11. How does water pressure affect cutting performance?
   A) Higher pressure reduces cutting ability
   B) Lower pressure increases speed
   C) Higher pressure increases cutting power and speed
   D) Water pressure has no major impact

12. What risk is associated with piercing brittle materials like glass or ceramics?
   A) Cracking or shattering during pierce
   B) Taper widening
   C) Abrasive clogging
   D) Excessive abrasive cost

13. Why use a 5-axis waterjet over a 3-axis one?
   A) To cut thicker materials
   B) To achieve beveled or angled cuts
   C) To reduce abrasive usage
   D) To lower machine cost

14. What is a typical abrasive flow rate for most waterjets?
    A) 0.1–0.5 lbs/min (45–230 g/min)
    B) 0.5–1.5 lbs/min (230–680 g/min)
    C) 2–5 lbs/min (0.91–2.27 kg/min)
    D) 10–20 lbs/min (4.54–9.07 kg/min)

15. What software feature can help reduce cycle times?
    A) Pierce delay increases
    B) Slowdown on straight cuts
    C) Optimized lead-ins and reduced rapid moves
    D) Decreased pump pressure

16. What safety measure is crucial when operating at 60,000 psi (≈ 4 100 bar)?
    A) Wear regular safety glasses
    B) Check all high-pressure lines and fittings regularly
    C) Reduce water temperature
    D) Turn off air assist

17. What is the main challenge of cutting laminated materials?
    A) Excessive noise
    B) Increased water consumption
    C) Excessive abrasive wear
    D) Delamination or layer separation

18. How does water hardness affect pump maintenance?
    A) Hard water causes scale and reduces pump life
    B) Hard water reduces maintenance needs
    C) It has no effect
    D) Soft water clogs the nozzle

19. What is the role of slurry management in waterjet shops?
    A) Reusing spent abrasive
    B) Reducing water pressure fluctuations
    C) Controlling spent water and abrasive disposal
    D) Increasing edge smoothness

20. How can you improve cut quality on delicate materials?
    A) Increase abrasive flow and slow speed
    B) Reduce pressure, slow down, and use fine abrasive
    C) Use larger nozzles
    D) Cut without abrasive

Black anodized aluminum cut on a CNC waterjet.

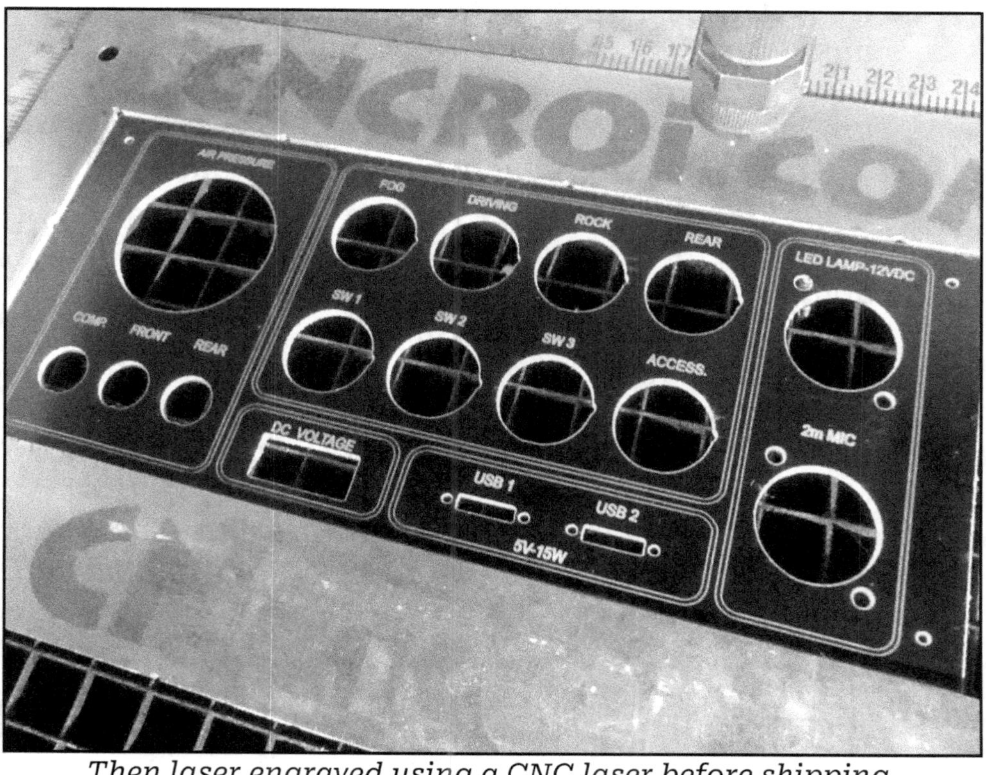

Then laser engraved using a CNC laser before shipping.

✖ 2.38 Answer Key and Explanations

This answer key cuts deeper than a waterjet on max pressure, but relax, the explanations are smoother than a taper-free edge.

1D – *Garnet is the most commonly used abrasive in waterjet cutting due to its hardness, availability, and low toxicity. It balances cutting efficiency with nozzle wear and is suitable for a wide range of materials.*

2D – *Kerf width (the width of the cut) is mainly influenced by the diameter of the nozzle and how far it is from the material surface (standoff distance). These determine the jet's focus and spread.*

3C – *Taper compensation adjusts the jet path to counteract the natural V-shaped taper created during cutting, especially in thicker materials, ensuring more precise vertical edges.*

4C – *Cutting too quickly doesn't allow enough time for the jet to fully penetrate thick material, leading to unfinished cuts or poor surface quality.*

5A – *High-pressure seals wear out over time due to intense pressure cycles. Replacing them regularly prevents leaks and maintains pump performance and longevity.*

6B – *The catch tank collects the high-pressure water and abrasive mixture after it exits the material, helping manage waste and reduce splashback.*

7A – *A proper standoff distance maintains cut quality. Too close risks impact damage to the nozzle, while too far leads to beam divergence and imprecision.*

8B – *Unlike laser or plasma cutting, waterjet cutting is a cold process. This eliminates the heat-affected zone (HAZ), preserving material properties and minimizing thermal distortion.*

9B – *Recycling garnet lowers operating costs and reduces the volume of abrasive waste requiring disposal, which is both eco-friendly and economical.*

10C – Thicker cuts typically show striations (wavy lines) near the bottom due to the jet losing energy. While the top edge is smoother, the bottom may show roughness due to taper and energy loss.

11C – Increasing water pressure raises the kinetic energy of the jet, allowing it to cut faster and through tougher materials with better efficiency.

12A – Brittle materials like glass can crack when first pierced due to sudden pressure buildup. Controlled pierce routines (e.g., low-pressure pierce) help mitigate this.

13B – 5-axis waterjets allow the cutting head to tilt, enabling complex bevels, countersinks, and 3D contours that are not possible with 3-axis machines.

14C – This is the typical abrasive flow rate for general waterjet applications. Lower or higher rates are used for specific materials or cut quality needs.

15C – Efficient toolpaths that minimize unnecessary movements and optimize entry points significantly reduce total cycle time without sacrificing quality.

16B – Operating at 60,000 psi (≈ 4 100 bar) poses serious safety risks. Regular inspection ensures there are no leaks or weak points that could fail under pressure.

17D – Cutting laminated materials (like composites) can cause the layers to separate, especially under high pressure. Special techniques are used to avoid this.

18A – Minerals in hard water can lead to limescale buildup inside the pump and plumbing, increasing wear and maintenance needs. Water softening is often used.

19C – Slurry management involves filtering, settling, or recycling used water and abrasive to stay within environmental regulations and reduce disposal costs.

20B – Delicate materials require a gentler touch. Lower pressure, slower cutting speeds, and finer abrasives minimize risk of cracking or surface damage.

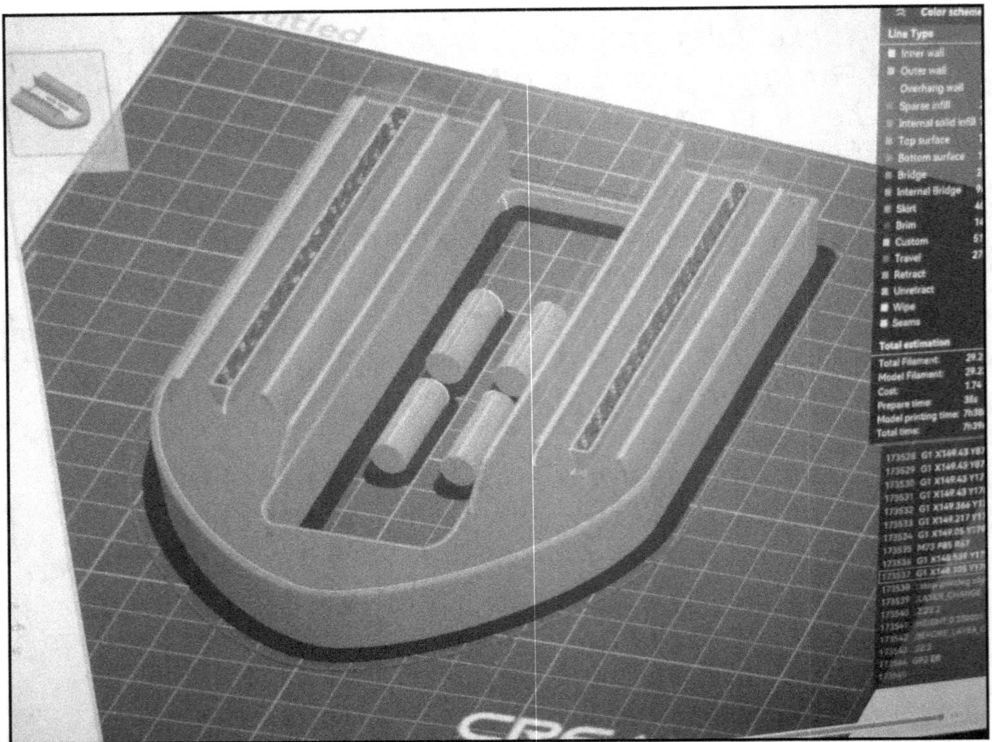

Three dimensional virtual design for FDM fabrication.

From digital design to a physical PLA part in just a few hours.

Section 3
The 30 Variants of CNC 3D Printing
Nozzle Knowledge & Bedside Manner

CNC 3D printing is basically the rebellious cousin of traditional CNC machines. Instead of cutting, carving, or blasting material away, it adds material layer by layer, weaving filaments into structure, like a sorcerer armed with an engineering degree.

Want to print a miniature spaceship? Done. Need a prototype for an oddly shaped part nobody's ever seen before? Easy. How about custom tooling, jigs, or even a replacement part for your coffee machine? Yep, a CNC 3D printer's got you covered, and like its CNC cousins, it won't complain about long hours, either.

At its heart, 3D printing (aka additive manufacturing) uses digital design files to guide the printer's nozzle or laser across X, Y, and Z axes, extruding or fusing materials like plastics, resins, metal powders, or even carbon-fiber-reinforced composites.

Think of it like a hot glue gun with a PhD in engineering: it knows exactly where to lay each drop or layer, precisely stacking them into functional parts, artistic creations, or high-performance components.

3.1 Why does industry love CNC 3D printers?

For starters, it smashes traditional manufacturing barriers. Instead of waiting weeks for custom tooling or molds, companies can 3D print prototypes or production-ready parts in a matter of hours. Need to iterate a design ten times in a day? No problem.

Traditional manufacturing says, "Let's set up the line." 3D printing says, "Hold my filament." This speed and flexibility empower industries like aerospace, automotive, medical, and even fashion to innovate faster and smarter.

Another game-changer: complexity is free. With CNC milling or turning, the more complex your part, the more expensive and time-consuming it becomes. But with 3D printing, complex geometries, internal channels, lattice structures, and organic shapes can be printed just as easily as simple blocks.

This means engineers and designers can optimize parts for weight, strength, or function without worrying about how to machine them afterward. It's like giving your design team superpowers. Think: the resolution of a laser, without needing any hold-downs like you'd use for a router.

And let's not forget materials. While most people think of plastic filament printers (the desktop FDM machines), industrial CNC 3D printers work with metal powders, engineering-grade resins, and even exotic materials like PEEK or Ultem.

You can print lightweight aerospace brackets, biocompatible medical implants, or heat-resistant engine components, all without expensive tooling or long lead times.

3.2 3D Printers: The Friendly Face of CNC

Is a CNC 3D printer the best "first machine" for everyone entering the CNC world? Maybe not. They can have a steep learning curve when it comes to 3D modeling.

They are undeniably fun and packed with potential. Plus, nothing beats the magic of watching a machine slowly build up a part from thin air while you sip your Irish coffee, pretending you're overseeing a futuristic microfactory.

In short, CNC 3D printing is the ultimate tool for rapid innovation, custom manufacturing, and creative experimentation. It's changing how industries design, prototype, and even produce final products.

So if you want to join the CNC revolution with a machine that doesn't just cut, but creates, the future is printing, and it's only getting better from here.

3.3 3D Printer Safety Myths & Mistakes

Welcome to the wonderfully warped world of 3D printing, where dreams and nightmares are built layer by layer.

Whether you're a seasoned printer whisperer or a fresh-faced filament feeder, you've probably fallen for at least one of these classic myths or made a mistake so gloriously dumb it deserves to be preserved in PLA.

Myth #1: "It's just plastic, how dangerous could it be?"

It's just plastic... like lava is just warm dirt. When you heat up thermoplastics, especially the fancy ones like ABS or nylon, they release chemicals that smell like what Dr. Doom would have farted during a fight with Spiderman.

Inhaling that stuff isn't something your body should have to adapt to over time.

Sure, PLA's the granola of filaments, but even it sneezes out ultrafine particles. So no, printing in a closet without ventilation isn't "efficient," it's DIY gas chamber vibes.

Get an air purifier, open a window, or at least stop acting like you're printing inside a submarine.

Myth #2: "PLA is totally safe indoors."

PLA is the "harmless" cousin at family barbecues, until he burns something down with a Roman candle. Yes, PLA is corn-based and smells like waffles when it prints, but don't let that fool you. It still releases tiny particles that your lungs definitely didn't sign up for.

Starting to see a theme here? Like other CNCs (and fabrication in general), it's what you don't see that kills you over time. Even I've cut my own lifespan in this business.

Just because it doesn't make your eyes water doesn't mean it's giving your lungs a hug. Ventilation is still your best friend. Or worst case, a good air filter, not the one covered in three months of cat hair.

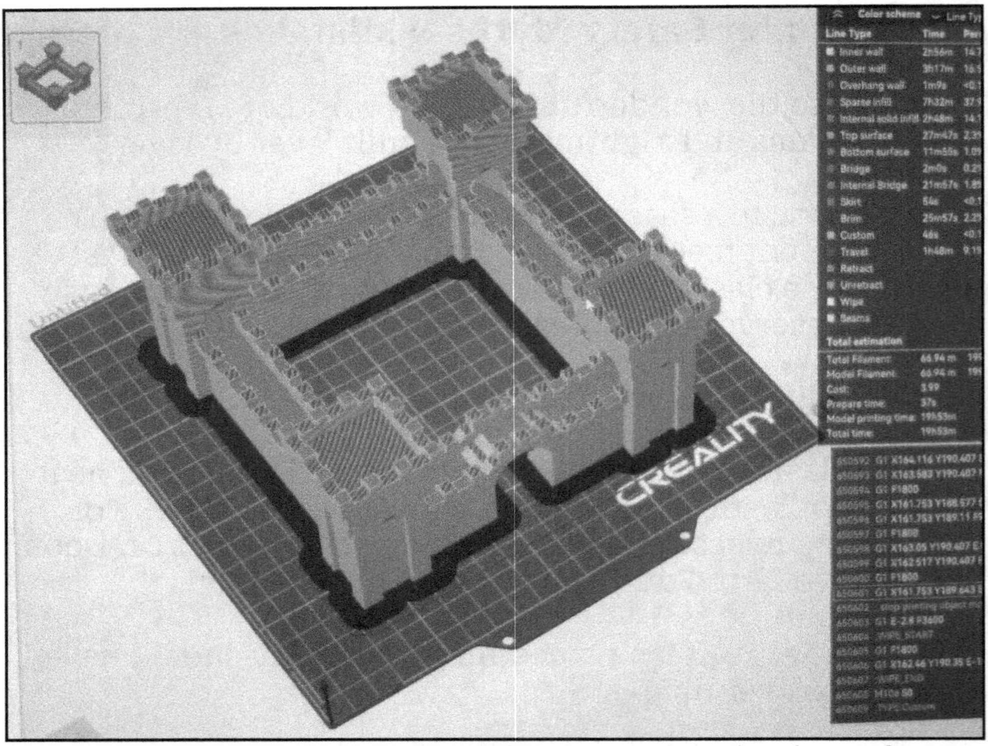

A decade of major breakthroughs in 3D pre-visualization software.

Myth #3: "I can leave it running overnight."

Leaving your printer running while you sleep is like deep-frying bacon in bed. Sure, it's probably fine... until it's very much not. Fires, spaghetti monsters, and motor meltdowns all love the midnight shift. One minute you're dreaming of perfect layer lines, the next you're explaining to firefighters what a Benchy is.

Unless your printer has industrial-level safety features or you are married to a smoke detector, consider babysitting it or installing a fire suppression system. At the very least, stop pretending your 3D printer is a crockpot.

Myth #4: "I don't need gloves for post-processing."

Between sharp support structures, resin goo, and that mysterious hobby knife you found under the couch, your fingers are in danger. Remember, there is such a thing as chemical accumulation in your body and the chemistry experiment your 3D printer is doing is part of that.

Gloves aren't just a fashion statement; they're the difference between "cool print" and "urgent care bills" possibly now and well into the future. Whether you're sanding, scraping, or dipping things in solvents you can't pronounce, glove up like you're about to enter a sci-fi airlock run by HAL 9000.

Myth #5: "The enclosure is just for print quality."

Yes, enclosures help with warping, but that's like saying walls are just for acoustics. A good enclosure protects you from heat, fumes, and the urge to poke moving parts while yelling, "Is it done yet?" It also keeps curious pets, kids, and sleep-deprived adults from getting too close.

That acrylic box isn't a status symbol, it's a safety device in disguise. Think of it as your printer's panic room: keeps the danger in, keeps the chaos out. Plus, it's the only time you'll see something benefit from being inside a plastic box.

Myth #6: "Resin is safe once cured."

Technically true, but only after it's fully cured and all the toxic slime has been banished. Until then, resin is like a passive-aggressive roommate: looks harmless, ruins everything. Uncured residue can sneak onto surfaces. And no, that thin tacky film left after washing isn't "extra shine," it's a skin irritation party waiting to happen.

Treat all your prints like they're radioactive until post-cure. And wash your hands like you just handled a ghost pepper.

Myth #7: "Electrical mods are easy."

You watched a 12-minute YouTube video and now you think you're Tony Stark with a soldering iron. Cool story. But swapping out random PSUs, hotends, or installing that sketchy RGB mod from a forum post could turn your beloved printer into a smoke machine.

Electricity doesn't care if your wire gauge is "close enough." Poor wiring = fire hazard. At best, you pop a fuse. At worst, your print room becomes a Reddit thread.

Myth #8: "I can clean it with whatever's handy."

WD-40? Sure. Acetone? Maybe. That mystery cleaner from under the sink? Hard no. Your printer isn't a toddler's high chair, random sprays and household cleaners can ruin electronics, crack plastic, or start fires when applied to hot components. Ever seen isopropyl alcohol on a heated bed? Boom, accidental mini-flamethrower.

Always check compatibility, read labels, and for the love of melted extruders, don't clean with flammable stuff while the machine's on.

Myth #9: "It's just plug and play."

Technically, yes, if you define "play" as six hours of bed leveling and Z-offset tuning. Even the most beginner-friendly printer demands calibration, firmware checks, and a gentle touch with a hex key.

Treating it like a toaster will just get you burnt toast.

"Plug and play" is a phrase invented by marketing teams who've never dealt with thermal runaway or Y-axis binding. Read the manual. Level the bed. Respect the nozzle. This isn't a toy, though it can make them; it's a high-temperature robot that sometimes eats itself.

Myth #10: "More speed = better prints."

Cranking the speed is fun until your printer starts sounding like a router spindle in need of a rebuild and gives you a print that looks like it was drawn during an earthquake. Yes, speed matters, but so do stability, cooling, and not launching your build plate into the wall.

Faster isn't always better, unless you like layer shifts and under-extrusion.

Your printer isn't in a race; it's doing delicate, molten ballet. So slow down, dial it in, and resist the urge to treat your slicer like a NOS injector. Remember, the tortoise won the race, not the hare.

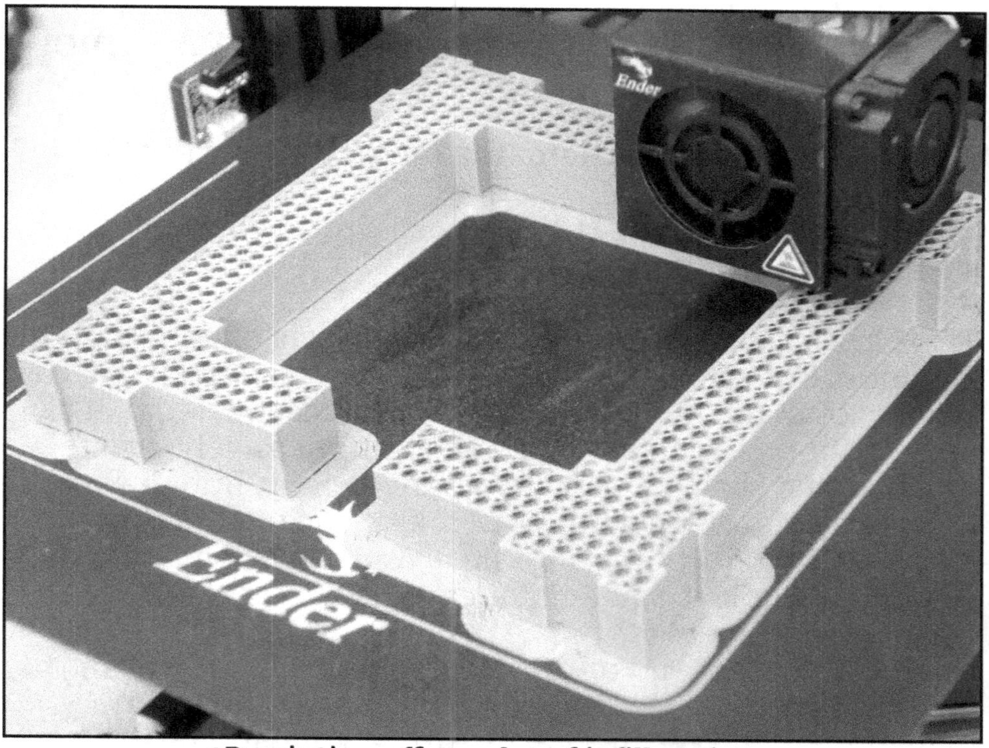

3D printing offers a lot of infill options.

3.4 Lessons I've Learned About 3D Printers

3D printers especially can turn imagination into reality, one tiny layer at a time. But behind the wizardry is a symphony of hot nozzles, shifting belts, spinning fans, and wires that, if neglected, will gleefully turn your weekend project into a miniature meltdown.

Skipping safety steps or falling for common myths can quickly transform a masterpiece into a mess. They've come a very long way since my 3D Systems Cube 13 years ago!

So like all the more dangerous looking CNC machines out there, wear your PPE, keep flammable cleaners away from hotbeds, don't wander off during 12-hour print marathons, and treat regular maintenance like a sacred ritual, not a side quest.

Because it's not just about printing cool stuff. It's about staying safe, staying smart, and being alive to admire your next beautifully over-engineered pencil holder.

☐ Spool Up Your Questions; It's Q&A Time!

3.5 How does nozzle diameter impact print resolution?

Think of it like picking the right pen tip, do you want to sketch fine art or graffiti a wall?

Tiny nozzles, like 0.2–0.3 mm, are the precision ninjas of the 3D world. They spit out ultra-thin yet consistent spaghetti strands of filament, letting you capture insane detail. It's perfect for miniatures, tiny gears, or impressively ornate dragons that seem to be everywhere on Etsy. But beware: these delicate widgets are slow. Like, "you could grow a beard waiting for that print" slow. And they clog up if you so much as look at a filled filament the wrong way.

Now on the other end of the nozzle spectrum, we've got the big boys: 0.6–1.0 mm. These chonky champs blast out filament like they're in a pasta-eating contest. They're fast, tough, and perfect for printing things like brackets, helmets, or full-size Iron Man suits (no judgment). But finesse? Not their strong suit. Expect chunky layers and surface details that scream "I was printed in a rush."

And let's not forget your slicer settings, they need to vibe with your nozzle size. You can't shove a flood through a straw or expect fine wine from a firehose. Pick a nozzle that matches your goals: high-def miniatures? Go small. Speedy big prints? Go large.

3.6 What is the effect of layer height on mechanical strength?

Too thick? Ugly. Too thin? Slow and possibly weak. Just right? Now we're talking.

A low layer height, say, 0.1 mm, is like printing with a precision razor. You get silky-smooth finishes and crisp details that make your print look professionally sculpted, like it belongs in a tiny art gallery. But there's a catch: it takes forever. Hobby? No problem. Business? Time to IPO.

Additionally, because the layers are so thin, they sometimes bond like shy introverts at a party, awkwardly and with limited commitment.

Crank up the layer height to 0.2–0.3 mm, and now you're in beast mode. Thicker layers mean stronger fusion between layers, especially in the Z-axis. Your parts get tougher, and your printer moves like it's got somewhere to be. But finesse? Out the window. Your print may end up looking like it was carved with a dull spoon by someone wearing oven mitts. Again, it comes down to learning your CNC machine and how your software outputs to your hardware.

And then there's the material. TPU (aka the wobbly noodle of filaments) plays by totally different rules than rock-solid PLA or grumpy ABS. What works like magic for one can be a total disaster for another. Thankfully, slicer software offers some guidance. Of course, each 3D printer, like any other CNC, has individual quirks to overcome as well.

So whether you're going for showroom shine or industrial strength, layer height isn't just a number, it's a choice that could make or break your print. Literally.

3.7 How do print orientation decisions affect part performance?

Since FDM prints are layer-cakes of plastic, they're weakest along the Z-axis. So, orienting load-bearing bits flat (in the XY plane) keeps them from snapping like dry pasta.

For looks, aim your best face up, fewer visible layer lines. But beware: this may summon the Support Goblins, leaving scars and post-print challenges. Orientation also affects warping, wobble, and whether your print topples like a drunk giraffe. With resin printers, it's even trickier as you have to deal with resin drainage, light curing, and funky overhangs all join the chaos.

In the end, orientation is a balancing act of strength, beauty, and print success. A bit of engineering, a bit of artistry and maybe a touch of sorcery. Keep exploring and expanding your skillset and I'm sure you'll figure it out.

3.8 How can you design better support structures?

Without them, your majestic print turns into a stringy mixed noodle heap the moment it tries to defy gravity.

Step one: avoid them like death and taxes. Reorient your part or redesign it to minimize overhangs, always because having fewer supports means less cleanup overall.

But when supports are necessary (and they will be), pick the right style. Grid? Strong but clingy. Linear? Old-school. Tree supports? Nature's gift to lazy makers, less material, fewer scars, and they snap off like overcooked twigs.

Tweak support settings like density and interface layers to hit that sweet spot between "holds firm" and "removes without rage." Or go full wizard mode: design custom supports in CAD that only show up where truly needed but that's generally reserved for engineering pros.

Got a dual-extruder printer? Use dissolvable supports like PVA or breakaway filament. It's like hiring a stunt double for your part, it does the hard work, then disappears without a trace other than the end credits.

And don't forget the endgame: removing supports without gouging your print. Plan for cleanup such as sanding, smoothing, or light sprays of acetone.

3.9 What are the differences between open-source and closed-source printers?

Open-source printers (like Prusa or my old Creality that's 40% zip ties) are the DIY playground of the 3D world. You get full access to hardware, firmware, and maybe even the slicer's soul. Tweak it, mod it, strap a laser to it, no one's stopping you. The community is huge, helpful, and slightly obsessed.

Bonus: your wallet will stay full. Downside? You're the tech support. When things go wrong, and they will as with any CNC, you'll be neck-deep in forum threads at 2 a.m., surrounded by hex keys and bleeding fingertips.

Closed-source printers (like Ultimaker or MakerBot) are the sleek, professional cousins in the 3D printing family. They show up in fancy boxes, smell like corporate approval, and usually just work right out of the gate. They politely request, well, demand, that you use their pre-approved filaments, slicers, and settings. The upside? Minimal stress, polished user experience, and customer support that actually answers the phone. Great if you are an elementary school teacher, not so much if you are a tinkerer.

The downside? A premium price tag and the creative freedom of a padded room. Want to swap out a hotend or flash custom firmware? Good luck, there might be digital handcuffs involved. Maybe even a firmware death glare. You get reliable prints, warranty peace of mind, and software that doesn't require you to Google "Why did my printer just scream?" Your call: Mr. MacGyver.

It's worth noting that Creality and Bambu Lab, once known for their open-source approach, have been shifting toward a more closed-source model, so things are very malleable in this space. Stay tuned folks!

3D printing delivers precision, even for very small items.

3.10 How do you prevent thermal runaway in printers?

It's not just a fancy term. It's when your printer's heater (usually the hotend or bed) decides to ignore all rules, crank up the heat, and head straight for meltdown mode. Best case, you get a toasted print. Worst case, your printer becomes a very expensive firestarter.

The main line of defense is a properly working thermistor, that little temperature sensor that tells your printer when to chill out. If it's loose, unplugged, or just lying about the temperature like a shady weather app, the printer may keep heating until the Sun is jealous.

That's where firmware safeguards come in. Smart firmware like Marlin includes thermal runaway protection. It basically acts like: "Hey, the heater's on, but the temperature isn't going up. Kill the power."

Don't stop there. Check your wiring, heater cartridges, and boards regularly. If your printer looks like it was wired by raccoons, it's time for maintenance. Consider adding thermal fuses or cut-off switches as backup systems because redundancy is sexy when fire is on the line.

Using an older or bargain-bin printer you got cheap on eBay? Flash it with updated firmware that includes protection. Seriously. Do it. Your house will thank you.

If you think this can never happen on a production machine, think again. Early batches of Creality's H2D printers had power supplies wired for 220V by default. If you plugged one into a 110V outlet without switching the voltage selector or if it was misconfigured from the factory, the components might not get enough power to function properly. This could cause thermal stress or erratic heating behavior, exactly the kind of thing that could lead to thermal runaway if safeguards fail.

Unless you are printing in a concrete bunker with a fire extinguisher and no emotional attachment to the building.

Never run any CNC machine unattended. Ever.

3.11 What's the importance of filament drying for hygroscopic materials?

Some filaments, such as nylon, PETG, and TPU, are hygroscopic, which is a fancy way of saying they suck moisture out of the air like a thirsty sponge, just like MDF. And when you try to print with that damp noodle? Yikes. You'll get stringing, bubbling, weak layers, and surface textures that look like your printer sneezed mid-print.

Here's what's really happening: that absorbed water turns into steam inside your hotend, causing tiny plastic burps (a.k.a. micro-explosions) that ruin extrusion consistency and make your print look like it survived a minor volcano eruption.

The fix? Dry it out. You can use a fancy filament dryer, a food dehydrator, or even your oven (just watch it closely as you're drying filament, not roasting it). Ideal temps are material dependent.

Once dry, keep your spools in airtight containers with desiccant, unless you want to repeat the whole drying drama in a week. Pro shops even use heated dry boxes that feed filament straight into the printer, like a perfectly controlled IV drip of plastic goodness. Most manufacturers sell these as options with their 3D printers nowadays.

3.12 How does multi-material printing affect slicing strategies?

Mixing filaments like PLA with PVA or TPU lets you print in color, use dissolvable supports, or combine flex and rigid parts. But your slicer has to manage tool or material changes with purging depending on your set-up.

Each switch usually means a purge tower or prime line, adding print time and waste. Rebranded industry standards like Orca Slicer let you tweak purge settings, transition speed, and tower placement to minimize the mess.

Material matchups matter as different temps, adhesion, and shrink rates can clash.

Pre-visualization is almost 100 percent accurate.

Perfect on screen. Gravity approved in real life.

3.13 What about firmware upgrades?

Updating your printer's firmware can unlock all kinds of magic such as better prints, smarter features, and fewer "why is it vibrating like that" moments. Open-source options like Marlin, Klipper, and Repetier keep evolving like Pokémon, adding superpowers like:

- Linear advance: for smoother extrusion.
- Input shaping: fights printer wobbles like a ninja.
- Thermal protections: because fire = bad.

Upgrade your firmware and suddenly your printer can handle fancy auto-bed leveling with BLTouch or inductive probes, finesse its microstepping, and dial in movement like it's training for a ballet recital. Say (mostly) goodbye to ringing or ghosting.

Got a Raspberry Pi lying around? Klipper will gladly borrow it to handle motion planning like a boss, while OctoPrint gives you remote control, webcam views, and the ability to pause prints from the comfort of your couch.

BUT WAIT. Firmware isn't a casual update (for any CNC). It's not a Windows patch. Flash the wrong version and you could brick your printer into silence. Always check compatibility, back up your config, and follow guides like your life (or at least your Benchy) depends on it.

By now, I really hope you're picking up on the recurring theme in this book (and, let's be honest, my last one too): CNC machines may come in different flavors, but the underlying problems can be eerily familiar.

Take firmware updates, sure, we're talking 3D printers here, but it's the same high-stakes game with plasma cutters, waterjets, or lasers. One wrong move and boom: congratulations, you've turned your machine into an expensive paperweight.

That's why, at *CNCROi.com*, I follow one golden rule: if it ain't broke, don't flash it. Unless an overpaid certified tech is there to pick up the pieces.

3.14 How does enclosure temperature affect large prints?

When you're printing large parts with high-temp materials like ABS, ASA, or polycarbonate, an enclosure isn't optional. Without it, uneven cooling leads to warping, cracking, and corners that lift like they're trying to escape.

Enclosures trap heat, letting prints cool slowly and evenly, kind of like tucking your print into a warm blanket so it doesn't freak out halfway through. This helps layers stick together and keeps shrink-happy materials in check.

But don't overdo it with the warmth and comfort. Too much heat and things get weird. PLA turns into a noodle, electronics start sweating, and tiny details suffer. That's why fancy printers have ventilated or temperature-controlled chambers to keep the climate just right automatically.

Even a DIY box made from IKEA leftovers can make a massive difference. Bottom line: if you're scaling up, enclose it.

3.15 What are the benefits of linear rail systems over rods?

Unlike basic smooth rods and bearings (which work fine until they don't), linear rails like MGN or Hiwin use hardened steel blocks that glide with precision and swagger. They're stiffer, smoother, and way less prone to wobble, flex, or sag. This is especially true during fast moves or big prints. The bigger the CNC, the more problems you'll have with perfect XYZ coordination by default.

Smooth rods can wear unevenly, misalign easily, and eventually turn your printer into a shaky mess. Linear rails? They spread the load evenly, handle high speeds like a champ, and keep prints sharp and clean.

Yes, they cost more and need precise installation (mess it up and you've built a very expensive rattle), but for high-speed or large-format printing, they're a serious upgrade.

3.16 How do you optimize slicer settings for TPU?

It's durable and impact-resistant, but printing TPU is more ballet than drag race. Slow it down to avoid pressure buildup that causes blobs or under-extrusion.

Retraction? Easy does it. Go for short and slow pulls, as too much causes jams. Too little and it strings like it's decorating a cake. Turn off or tone down coasting and combing while you're at it.

Flow rate needs tuning, TPU extrudes differently than rigid filaments. Dial in your extrusion multiplier to avoid globs or gaps. Stick to moderate layer heights for good bridging and layer bonding.

Bed adhesion's usually great on PEI or textured surfaces, but a bit of glue stick helps with bigger prints.

With the right settings and some patience, TPU gives you prints that bend, bounce, and survive a beating. Again, just don't rush it to the finishing line and you'll be fine.

When speed matters, routers can outperform 3D printing.

3.17 What are the key limitations of direct-drive extruders?

In a direct-drive setup, the extruder motor sits right on top of the hotend like a tiny, overly enthusiastic backpacker. This gives you amazing filament control, especially for squishy stuff like TPU or TPE. That's why brands like Prusa, Bambu Lab, and Creality's higher-end models love them. They're basically the go-to for flex-filament finesse.

But there's a catch: that motor's not weightless. Slap it on the moving printhead and now your machine's trying to do high-speed ballet with ankle weights. The result? Lower max speeds, more vibration, and surface artifacts like ringing that make your print look like it's echoing.

Also, maintenance is... let's say "intimate." Want to clear a clog? Hope you enjoy unscrewing 12 things in a space the size of a Tic Tac box. Tight integration can make swaps and repairs a puzzle. Things have gotten a lot better though, even since the release of my first book in this series.

On the plus side, the shorter filament path is a dream for flexibles, but it also means brittle filaments (like carbon-fiber blends) can snap if they're in a bad mood.

Cooling and part access? Slightly trickier, since you're designing around a bulkier, front-heavy printhead. But if you want versatility and superior extrusion control across a variety of filaments, direct drive still reigns supreme, as long as you don't mind a little extra weight up front.

Think of it as the sports car of extruders: fast, nimble, and high-maintenance, but totally worth it when tuned right.

3.18 How can you integrate sensors for real-time failure detection?

These days, most modern printers come loaded with sensors right out of the box because nobody enjoys waking up to a 12-hour spaghetti sculpture where your print was supposed to be. Recently, 3D printers have made significant strides in both ease of use and overall capabilities.

Filament runout sensors are the MVPs of basic failure prevention. They detect when your spool runs dry and pause the print so you can reload, instead of letting your printer air-print into sadness for hours.

Then there are filament flow sensors (a.k.a. "is-this-stuff-actually-moving" sensors). If your filament gets tangled, snapped, or clogged, they stop the show and raise the alarm, saving you from a molten mess and wasted filament.

Auto-bed leveling sensors like BLTouch and inductive probes are now pretty standard too. They save you from endless manual bed tweaking and dramatically reduce first-layer freakouts. 3D printers are getting exponentially better!

And if you're really fancy (or mildly paranoid), some setups include accelerometers, piezo sensors, or even machine AI vision, yes, literal cameras, to detect crashes, layer shifts, or spaghetti storms in real time. Pair that with OctoPrint and your printer can send you texts like, "Hey, something's wrong, and I'm scared."

3.19 What's the best way to handle warp-prone filaments like ABS?

This stuff loves to warp thanks to its high thermal shrinkage, so step one is locking it in a heated enclosure (even a DIY box helps). Keep the ambient temps cozy to prevent it from curling up like a stressed potato chip. Got an AI controlled heated chamber? Even better.

Bed adhesion is everything. ABS clings best to PEI, Kapton, Magigoo, or old-school ABS slurry (a.k.a. ABS + acetone soup). And don't even think about skipping the heated bed unless you enjoy watching corners lift like they're trying to escape. Add brims or rafts to anchor it down, and leave your part cooling fan off, or low, so it doesn't freeze mid-print. ABS hates cold breezes even more than PLA.

Design tricks help too: round off sharp corners, add fillets, or slap on some "mouse ears" (tiny sacrificial tabs) to keep edges grounded. Trial and error, software through hardware, is the best way forward as always.

3.20 How do you calculate print time from gcode files?

Slicing software like Orca Slicer and PrusaSlicer give you time estimates based on toolpaths, speeds, and layers. It's helpful but not exactly gospel. These estimates don't factor in things like firmware quirks, acceleration limits, or the moment your printer pauses mid-layer to question its existence.

For better accuracy, tools like OctoPrint analyze actual G-code behavior on your specific machine. Newer printers often show estimates too, though they still lie, just a bit more politely like all the other CNC machine platforms.

Long print? Expect drift. Like with big CNC jobs, the longer the run, the more things veer off. A 10-hour estimate can easily turn into 12, or "surprise, it's still going tomorrow."

You can calculate time manually from gcode... if you're into pain. Otherwise, let the software guess, and pad your estimates. No one ever said, "Wow, I wish this had taken longer." ETA should be taken with a grain of salt.

Even cheap 3D printers can produce clean bottom surfaces.

3.21 What post-processing methods improve resin print durability?

Fresh off the bed, SLA/DLP prints may look amazing, but they're basically brittle little delicate showpieces until properly post-processed. Things have gotten way better though as they gain more mainstream momentum with the general public that isn't interested in tinkering.

Step one: wash the goo. Use isopropyl alcohol or a wash station to remove uncured resin. Skip this, and you'll have sticky prints that smell weird and snap under pressure.

Step two: hit it with UV. Prints aren't fully cured until they get a solid UV bath; either in a fancy curing chamber or under a UV lamp (sunlight works too in a pinch). Curing boosts strength, hardness, and chemical resistance. Don't worry too much about over-curing BTW.

Some folks go further, such as clear coats, sanding, or even sealants to toughen up fragile bits or smooth sharp edges. Resin still won't beat ABS in a cage match, but it'll survive more than one passive-aggressive stare.

Post-processing isn't just for resin, FDM prints benefit too.

3.22 How does the choice of infill pattern affect stiffness?

Infill patterns like grid, cubic, and gyroid spread forces like champs, making your parts stiff, strong, and ready for action, perfect for load-bearing stuff. Want to save time and filament? Lines or zigzag print faster, but they're the limp noodles of structural support.

Honeycomb hits a sweet spot: lightweight and decent strength, but prints slower thanks to its maze-like travel paths. 10–20% infill is fine for looks; 40–60% for strength. More than that? You're just showing off (and wasting filament) unless your application (or client) calls for it.

Pick infill based on strength, speed, or print vibes. Smart choices mean less waste and fewer surprise collapses.

3.23 What are the differences in mechanical properties between SLA and FDM?

SLA prints are the runway models of 3D printing, when working right, they have flawless skin, killer detail, and strength in all directions. Each layer cures as one solid slab, so you get smooth parts with great tensile strength and barely a layer line in sight.

But don't let the looks fool you, standard resins can be brittle. One good flex or drop and snap! Specialized resins can toughen things up, but they'll cost you, and UV light still makes them age like milk if you skip proper post-processing.

FDM, meanwhile, is your blue-collar bruiser. Materials like PETG, ABS, and nylon shrug off daily abuse, flex, and impact like it's no big deal. The catch? They're stronger sideways than upwards as layer adhesion is their Achilles' heel. Oh, and they're not winning any beauty contests without sanding and paint.

3.24 How do you manage resin disposal safely?

Uncured resin is hazardous waste, not smoothie material. Like cooking oil, never dump it down the drain.

Instead, cure it solid. Hit leftover resin or messy paper towels with UV light (sunlight works, but UV lamps are faster) until it's a hardened, non-toxic lump.

Once cured, most places let you toss it as regular trash, but check your local rules before playing trash roulette.

Used IPA? Let the resin settle, pour off the clean stuff to reuse, and cure the leftover sludge before tossing. Bigger setups often hire hazardous waste pros to deal with it properly.

Always wear gloves, eye protection, and store resin in sealed, labeled containers, preferably somewhere that won't turn it into surprise slime if sunlight hits.

3.25 How do flexible build plates improve part removal?

These spring-steel, magnet-backed wonders have revolutionized print removal. Back in the day, getting a print off glass or aluminum meant chiseling, prying, or straight-up rage. Risked your fingers and your print. I still have PTSD from my 3D Systems Cube glass plate.

Now? Just let the bed cool, pop off the plate, and flex it like a credit card, the print usually pops right off. No tools or broken parts. They do wear out though, unlike glass but you're still ahead of the game in my book.

Bonus: many come with PEI or textured coatings that grip during printing but let go once cooled. They also swap out easily for cleaning or surface changes, which is great if you print often or mess up often. Always have a backup around.

3.26 What are the best practices for dual extrusion calibration?

Printing with two materials opens up cool stuff, multi-color prints, soluble supports, flexible-rigid combos.

Step one: nozzle offset calibration. If your nozzles aren't perfectly aligned in X, Y, and Z, you'll end up with a dual-extruded disaster. Most slicers help with this, or you can print fancy calibration doodles and fine-tune manually.

Step two: temperature juggling. Different filaments = different needs. Your slicer has to coordinate bed and nozzle temps like a dance routine. Mess it up, and you'll get blobs, stringing, or hot nozzle soup. Use prime towers, wipe walls, or ooze shields to manage the mess.

Retraction and wipe settings? Also need love. Your single-material profile won't cut it, tweak those or prepare for filament boogers.

Printing with soluble supports like PVA? Keep that stuff dry or it'll feed like soggy noodles. Oh, and don't rush. Dual extrusion may take longer and require more tweaking, but once it's dialed in? You're awesome!

3.27 How does print cooling affect overhang performance?

In FDM printing, especially with overhangs and bridges, cooling fans are your plastic's personal trainers. It solidifies molten goo before gravity ruins everything.

With materials like PLA, crank the fan up as aggressive cooling means cleaner angles and sharper details. But with ABS or PETG, too much cooling can cause warping, cracking, or prints that peel like old wallpaper.

Slicer tweaks help: slower speeds on tricky layers give fans time to work, and smaller layer heights reduce saggy step sizes. Or get fancy with chamfers or built-in supports to give gravity a break.

The secret? Cooling isn't "blast fan = win." It's a balance to match the airflow to the material and design, and your print goes from "slouchy disaster" to "chef's kiss."

3.28 What are key differences between Bowden and direct-drive setups?

Once upon a layer, in the magical land of 3D printing, two extruders argued over who was the real MVP.

Bowden, the lightweight speedster, puffed up his frame-mounted chest.

"I'm sleek, I'm fast, and I don't need to carry extra weight on my hotend. I've got a Bowden tube, which is basically a filament slip-n-slide! Less vibration, more speed, baby!"

Direct-Drive, the heavyweight control freak, rolled his gears and said:

"Cool story, Spaghetti-Tube. But let's talk control. I'm right there on the hotend, gripping filament like it owes me money. You think TPU's going to behave in your wobbly waterslide? Please."

Bowden smirked.

“Yeah, but you carry your motor around like an angry backpacker. Enjoy those ringing artifacts when you try to sprint.”

Direct-Drive shrugged.

“Fine, I’ll go slower, but I’ll do it clean, with flexibles, exotics, and whatever weird filaments the humans dream up next. You choke on TPU like it’s a noodle in a vacuum cleaner.”

The Verdict?

If you like speed and don’t mind being picky with filaments, Bowden’s your guy, but if you want to print flexible unicorn rubber with surgical precision, Direct-Drive’s your ride-or-die.

As with everything else, there is a constant evolutionary fine tuning occurring to make these differences less painful.

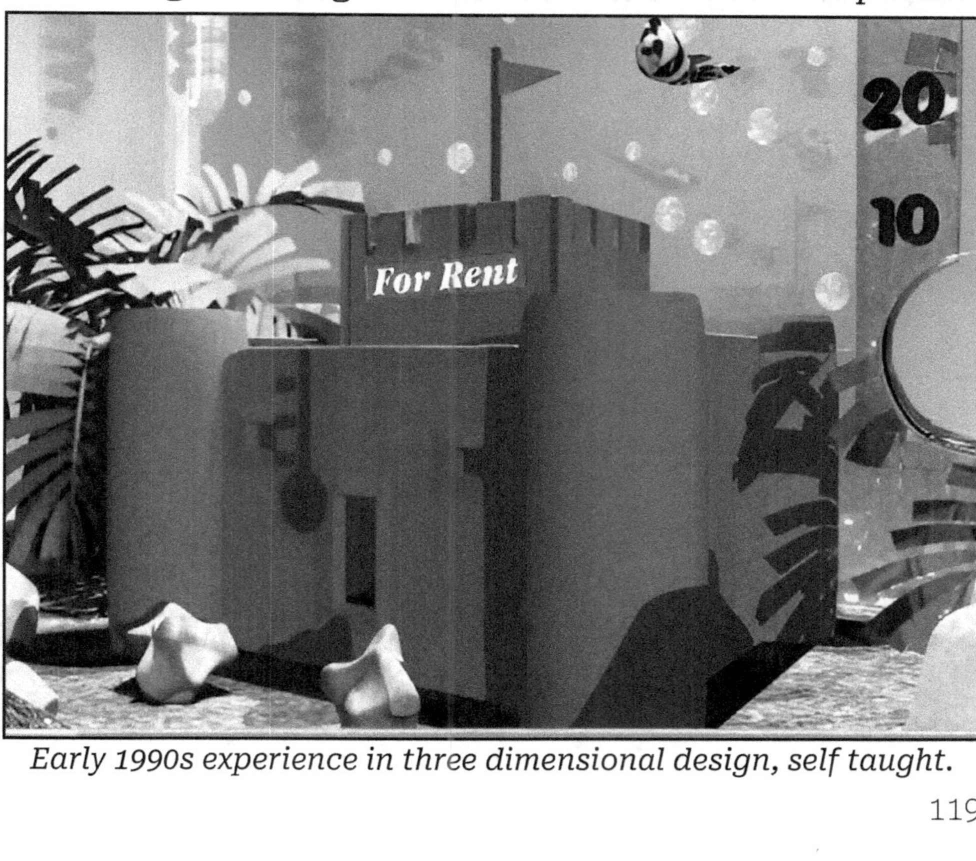

Early 1990s experience in three dimensional design, self taught.

My early days of 3D modeling were an exciting time to create.

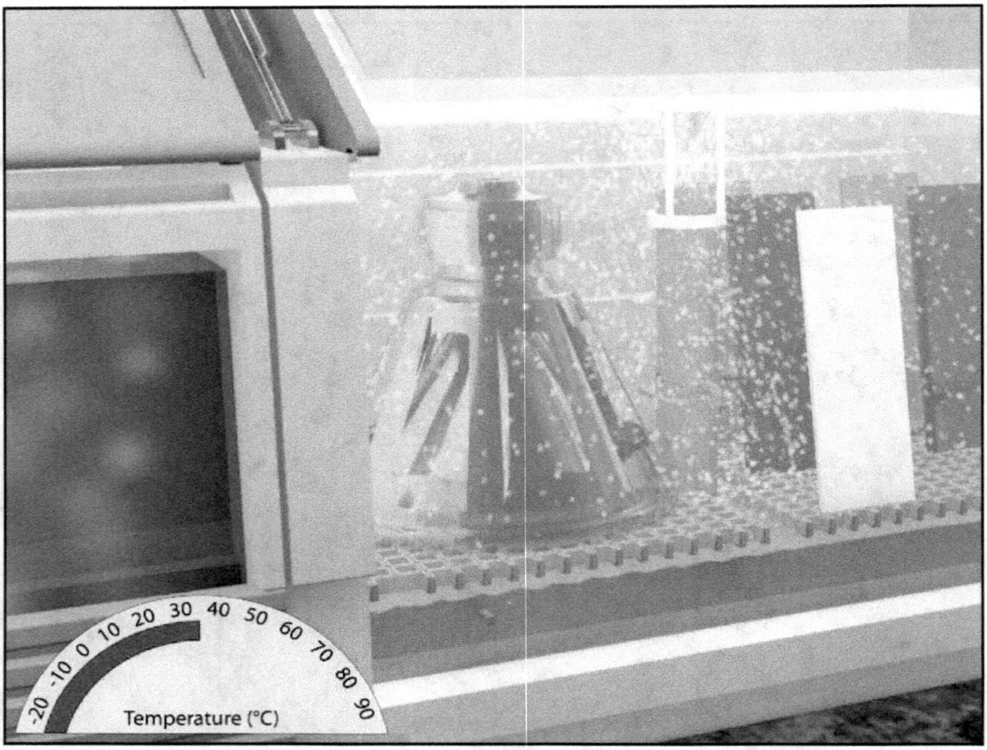

The same skills still apply to 3D printing today.

3.29 How do you tune retraction settings to prevent stringing?

You know those annoying little plastic hairs your print leaves behind, like it just walked through a haunted house? Yeah, that's stringing. It's what happens when your printer decides it's Spider-Man during travel moves. The culprit? Retraction settings gone rogue. You see this all over the tops of my 3D printed models using my Ender-3 Pro.

Retraction 101: It's your printer's way of saying, "Hold up, let me suck that back in," before moving from one spot to another. Think of it like wiping your mouth before leaving the dinner table, polite and less messy. Start by dialing in the retraction distance.

For direct drive systems, a retraction distance of 1–2 mm is usually enough while Bowden setups need more effort, typically 4–6 mm, thanks to their long-distance relationship with the extruder. Set it too short and you're in Cobweb City; too long, and you've clogged your printer like a toddler stuffing spaghetti into a trumpet.

Then there's retraction speed, the sweet spot depends on too many variables but it's usually 25–50 mm/s, also known as the Goldilocks zone: not too slow, not too fast. Go too slow and the filament dribbles like it's drooling; go too fast and your extruder throws a mechanical tantrum.

Temperature? Oh yes, it matters. Too hot and your filament turns into lava, oozing out like it's got no self-control. Lower it by 5–10°C (9–18°F) to give it some chill.

Travel speed is the art of ghosting: the faster the move, the less time it has to leave behind a mess. Be quick, be clean.

Feeling fancy? Enable coasting (stop early like a boss) or wipe (a tiny self-cleaning swipe). These features are like your printer's version of turning around to double-check it locked the door.

Still haunted by strings? Print a retraction tower or better yet, change to another material supplier.

3.30 What slicer settings improve bridging performance?

Bridging is when your printer tries to lay filament across open gaps like it's walking a tightrope with zero safety net. It's basically telling gravity to shove off while it tries to print mid-air. Predictably, it doesn't always go well.

Want to stop your bridges from turning into sad, droopy spaghetti? Let's fix that.

First up: speed. Slow it down. Think of your filament like a nervous gymnast, the slower it moves, the better chance it has of gracefully landing the stretch. Most slicers have a "bridge speed" setting that slows things down just for these dramatic moments.

Next: cooling. Blast that part cooling fan like it's summer and your filament just ordered a piña colada. Fast cooling means the plastic hardens in mid-air instead of sagging like a hammock under a walrus. Great for PLA, not so much for ABS, they're more like grumpy uncles who hate AC.

Now for extrusion. Less is more. Too much filament and it droops like an overstuffed lasagna noodle. Reduce the flow or extrusion multiplier during bridges. Many slicers already do this automatically, thinning the strand so it tightens like a proper bridge cable instead of flopping like a wet shoelace.

Finally, don't push your luck. Most printers can handle gaps up to 10–20 mm before the bridge becomes a suspension of the laws of physics.

If you're trying to span the Grand Canyon, either add supports or rethink your life choices (or at least your model design).

With a bit of slicer wizardry and some common sense, you'll be stringing up beautiful, sag-free bridges like a plastic-spitting architect in no time. Keep in mind that using dry and consistent (not the cheap stuff) filament also leads to fewer problems with 3D printing.

3.31 How does mechanical bed leveling differ from auto-bed leveling systems?

Manual leveling is the classic workout: twist the bed's corner screws while sliding paper under the nozzle like you're testing its grip strength. It's simple, cheap, and surprisingly effective until you bump your printer or sneeze near it and have to do it all over again.

Auto-bed leveling (ABL) is the future: a sensor pokes around the bed like it's scanning for treasure, then builds a magical compensation map to keep your nozzle at the perfect height. Some systems are faster than others.

ABL makes first layers way more reliable, especially on big or slightly warped beds. Just don't skip basic manual leveling entirely as the robot still appreciates a decent starting point, especially as the models grows in height.

Want fewer headaches and better prints? ABL is your new best friend, among many other bells and whistles.

3.32 What are best practices for large-format 3D printing?

Big prints aren't just scaled-up, they're a full-on boss fight. Everything gets harder, slower, and more dramatic.

First: stability. Large printers flex like yoga pros. Check frame, belts, and gantry often as wobbles = disaster.

Next: thermal control. Big prints in ABS or nylon will warp unless enclosed or chambered. Heated beds need to stay evenly hot. Think cozy blanket, not a small toaster oven.

Slicer settings? Crank up infill and wall thickness. Use brims or rafts so your print doesn't peel off mid-job. Long print? Break it into chunks, less heartbreak if it fails.

Final tip: prepare for chaos. Power cuts, clogs, tangles as they have more time to make their presence known. It's a good idea to invest in a UPS and spares of most problematic pieces of tech for quick replacement.

3.33 How do heated chambers impact engineering-grade materials?

When printing with engineering-grade divas like ABS, ASA, polycarbonate, or fancy nylons, a heated chamber isn't a luxury, it's survival. These materials hate cold drafts and will warp, crack, or split if the ambient temp isn't just right.

Heated chambers keep things cozy while reducing thermal shock between the hot filament and cool air. The result? Better layer bonding, less drama, and parts that don't tear themselves apart from the inside.

Got carbon- or glass-filled filament? A warm chamber helps prevent delamination and boosts strength. But be warned: your printer's guts (motors, boards, extruder) aren't heatproof. They'll need fans, insulation, or active cooling.

PLA, on the other hand, hates heavy blankets. But for serious materials and serious jobs, a heated chamber turns your printer from hobby toy to factory beast.

3.34 How do you handle print pausing and resuming during long jobs?

Need a break mid-print? Most modern printers can pause and resume just fine, perfect for filament swaps, tweaks, or mid-job sanity checks.

Pause tips: Retract filament slightly to prevent ooze, and park the nozzle away to avoid leaving a blob. Resuming usually happens at the same Z-height. Most printers now handle that automatically.

Power outage? Some printers have recovery modes that pick up where they left off. Not foolproof, though; cooled beds can cause weak layers or warping. A UPS = stress relief. Manually pausing? Avoid cosmetic surfaces or thin walls; seams might show.

With good timing and a little caution, pausing becomes a smart tool, not a gamble.

3.35 CNC 3D Printer: Top 10 Takeaways

Nozzle diameter affects resolution: *Small nozzles print finer details with smoother surfaces; large nozzles print faster but leave rougher edges.*

Layer height influences strength: *Thinner layers improve visual quality and detail; thicker layers print faster and often increase part strength due to better bonding.*

Print orientation matters: *Layer lines act as built-in fault lines. Orient parts to minimize stress across layers, especially for load-bearing or functional prints.*

Support structures need smart design: *Strategic support placement and minimal use reduce print time, material waste, and post-processing headaches.*

Filament prep affects print quality: *Moisture causes bubbling, stringing, and weak adhesion. Always dry hygroscopic filaments like nylon, PETG, or PVA before printing.*

Multi-material printing requires careful slicing: *Different materials have unique temperatures, speeds, and flow rates.*

Enclosure temperature controls print success: *A heated enclosure maintains ambient temperature, reducing warping, and layer separation on large or high-temp prints.*

Firmware upgrades enhance performance: *Updated firmware can add new features, improve motion control, enable advanced settings, and boost overall reliability.*

Post-processing improves durability: *Resin prints need full UV curing. FDM parts benefit from sanding, annealing, or smoothing (like acetone vapor for ABS) to improve strength and finish.*

Bed leveling is crucial: *A properly leveled bed ensures even first-layer adhesion, prevents warping or failed prints, and helps every layer stick with confidence.*

⊗ 3.36 Section Quiz

This quiz is like a 3D printer: take the wrong path, and things are gonna get stringy fast!

1. What effect does nozzle diameter have on print resolution?
   - A) Larger diameter improves fine detail
   - B) Smaller diameter improves fine detail
   - C) Nozzle diameter doesn't affect resolution
   - D) Larger diameter improves layer adhesion

2. What happens when you increase layer height?
   - A) Prints take longer but are stronger
   - B) Layer height has no effect
   - C) Prints go slower and get rougher
   - D) Prints go faster but lose fine detail

3. Why is print orientation important?
   - A) It determines how much filament you use
   - B) It affects part strength, supports, and surface finish
   - C) It only matters for color prints
   - D) It doesn't affect anything

4. What is the role of support structures?
   - A) They hold the part down to the bed
   - B) They reinforce the filament
   - C) They provide temporary scaffolding for overhangs and bridges
   - D) They increase speed

5. Which is a key difference between open / closed-source?
   - A) Open-source printers only print PLA
   - B) Closed-source printers require filament drying
   - C) Open-source printers allow personal modifications
   - D) Closed-source printers can't be used for schools

6. What is thermal runaway protection?
   - A) A safety feature that shuts off heating if sensors fail
   - B) A system to speed up cooling
   - C) A setting that keeps the nozzle hot between prints
   - D) A method to boost bed adhesion

7. Why is filament drying important for materials like nylon or PETG?
   A) It prevents stringing, bubbles, and weak layers
   B) It improves bed leveling
   C) It increases print color
   D) It sharpens overhangs

8. What challenge does multi-material printing introduce?
   A) Faster print times
   B) Simpler slicing
   C) More complex slicing and tool changes
   D) Less need for support material

9. What's the purpose of firmware upgrades?
   A) To change slicer settings
   B) To improve performance, add features, or fix bugs
   C) To upgrade filament types
   D) To reset the machine's factory settings

10. Why is enclosure temperature important for large prints?
   A) It prevents warping by maintaining even temps
   B) It cools the nozzle faster
   C) It increases speed
   D) It reduces nozzle clogging

11. What advantage do linear rails have over rods?
   A) Lower weight
   B) Reduced noise only
   C) More precise, smoother motion
   D) Slower print speeds

12. How should you adjust slicer settings for TPU (flexible filament)?
   A) Increase retraction and slow down
   B) Disable cooling fan
   C) Increase speed and retraction
   D) Reduce retraction and slow down

13. What is a key limitation of direct-drive extruders?
   A) They can't print TPU
   B) They require Bowden tubes
   C) They reduce extrusion precision
   D) They increase weight on the moving head

14. How can sensors help with real-time failure detection?
   A) Monitor filament runout or bed leveling errors
   B) Boost extrusion rates
   C) Speed up firmware upgrades
   D) Improve slicer settings

15. What's the best way to handle warp-prone filaments?
   A) Use no enclosure and increase cooling
   B) Use a heated bed and enclosure
   C) Print as fast as possible
   D) Reduce nozzle temperature

16. How do you estimate print time from a gcode file?
   A) By dividing filament weight by nozzle diameter
   B) Measuring print bed area
   C) Using slicer estimates or printer display readouts
   D) Checking firmware version

17. Which post-process boosts SLA resin durability?
   A) Sanding
   B) UV curing
   C) Painting
   D) Freezing

18. How does infill pattern affect part stiffness?
   A) No effect
   B) Different patterns distribute forces differently
   C) Only solid infill matters
   D) Infill only affects weight, not strength

19. What's the difference between SLA and FDM mechanical properties?
   A) SLA is usually more brittle, FDM is more flexible
   B) SLA is more flexible, FDM is more brittle
   C) SLA has no heat resistance
   D) FDM has no surface finish

20. How can you safely dispose of resin waste?
   A) Pour it down the drain
   B) Mix it with water
   C) Allow it to cure fully and dispose as solid waste
   D) Burn it outdoors

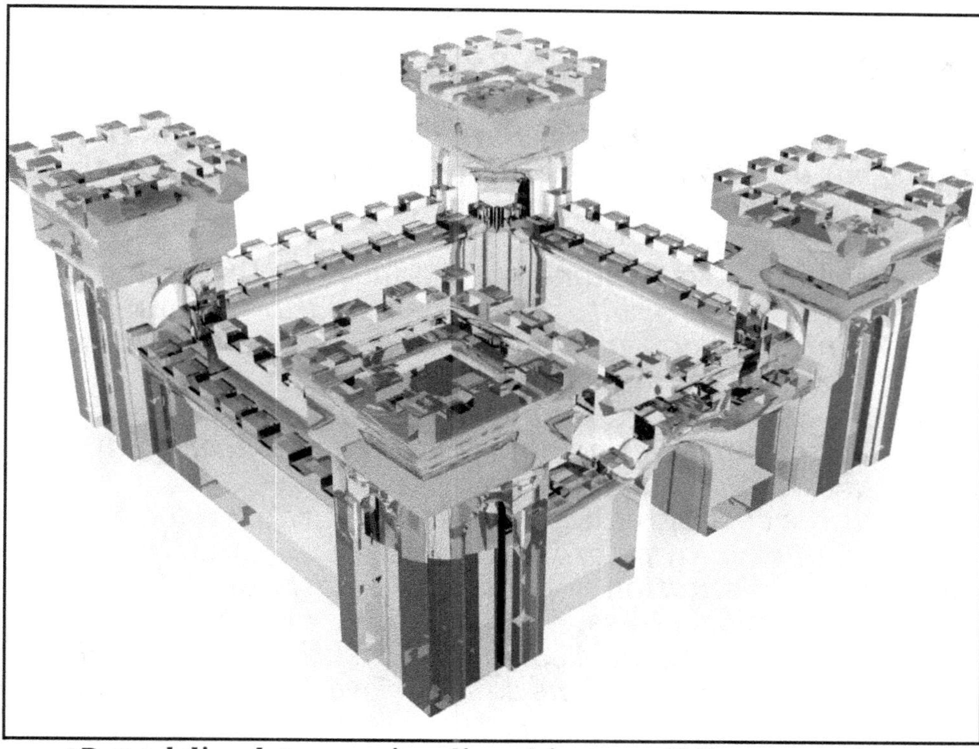

3D modeling lets you visualize objects as if they were glass.

If you are serious about 3D printing, learn 3D modeling.

✖ 3.37 Answer Key and Explanations

This answer key hits harder than a clogged nozzle mid-print, but don't worry, the explanations flow smoother than freshly tuned retraction settings.

1B – *Smaller nozzle diameters produce finer extrusion lines, which results in higher-resolution prints capable of capturing detailed features. While larger nozzles offer faster print times, they lack the precision needed for intricate geometry.*

2D – *Increasing layer height reduces the number of layers, allowing prints to finish faster. However, this comes at the cost of reduced detail and surface smoothness, as thicker layers are more visible.*

3B – *Print orientation impacts how a part handles mechanical stress, where supports are needed, and the finish of visible surfaces. Orienting layers to align with stress directions can improve strength and print quality.*

4C – *Support structures act as temporary scaffolding, holding up overhangs and bridges during printing. Without supports, sections of the print with steep angles or spans would collapse mid-print.*

5C – *Open-source printers allow users to modify hardware and firmware, offering flexibility and customization. Closed-source printers often restrict modifications to protect proprietary designs.*

6A – *Thermal runaway protection shuts off heating if the printer detects a temperature sensor failure. This prevents uncontrolled heating, which could lead to melted components or fire.*

7A – *Hygroscopic filaments like nylon and PETG absorb moisture, leading to bubbles, stringing, and weak layers during printing. Drying the filament ensures strong, clean, and accurate prints.*

8C – *Multi-material printing requires advanced slicing strategies and tool changes. Each material has unique settings, and switching materials mid-print demands careful coordination.*

9B – *Firmware upgrades can fix bugs, improve motion control, add new features like auto-bed leveling, or support hardware upgrades. Keeping firmware updated ensures optimal printer performance.*

10A – Heated enclosures keep ambient temperatures stable, reducing thermal stress in materials like ABS. This prevents layer separation and warping, especially in large or long-duration prints.

11C – Linear rails provide smoother and more accurate motion than rods due to tighter tolerances and less mechanical play. This leads to cleaner prints, especially at higher speeds or precision levels.

12D – TPU is soft and prone to buckling if retraction is too high or speeds are too fast. Reducing retraction and slowing down the print helps ensure steady extrusion and better print quality.

13D – Direct-drive extruders place the motor on the printhead, increasing its weight. This added mass can cause vibrations or limit print speeds, especially on fast-moving gantries.

14A – Sensors like filament runout detectors and bed leveling probes can catch issues early. These allow the printer to pause or adjust, preventing failed prints and wasted filament.

15B – ABS is prone to warping when cooled unevenly. A heated bed and enclosure provide a stable temperature environment, improving layer adhesion and preventing print deformation.

16C – Slicers and printer firmware estimate print time by analyzing toolpaths and speeds. These estimates are displayed after slicing or during printing and are more accurate than manual calculations.

17B – UV curing hardens SLA prints completely, improving their strength and durability. Without curing, prints remain soft and can degrade quickly, especially under stress or sunlight.

18B – Infill patterns like grid, cubic, or gyroid distribute stress differently within a print. Some patterns offer better strength in specific directions, affecting how rigid or flexible the part feels.

19A – SLA prints have a smoother finish and finer detail but are typically more brittle. FDM prints are tougher and more flexible, making them better suited for functional parts.

20C – Uncured resin is hazardous. It must be fully cured using UV light before disposal. Once solidified, it can be safely discarded as solid waste according to local regulations.

Even quarter inch steel slices instantly under plasma cutting.

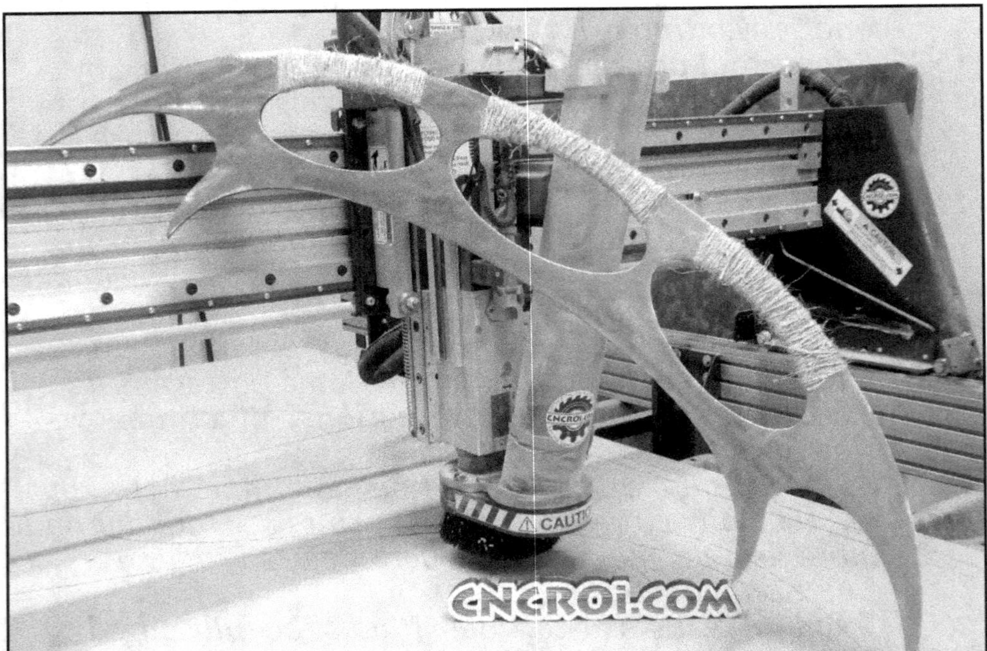

Custom weapons are fun to make in a CNC shop.

Section 4
The Plasma Table
Metal Melting with Precision

CNC plasma cutting is basically what happens when you hand a lightning bolt a map and tell it to follow a dotted line. It's loud, it's flashy, and it's one of the most satisfying CNC processes you'll ever be involved with.

Who doesn't want to command a controlled jet of superheated ionized gas (fancy talk for plasma) that slices through steel, aluminum, or stainless like voltage through an air gap? It's industrial firepower meets computer-guided accuracy, and it's been a game-changer for industries big and small.

So, what's really going on under the hood? A CNC plasma cutter takes a digital design file, usually a DXF or similar, and uses computer controls to move a plasma torch across a metal sheet along precise X and Y axes. Unlike mechanical cutting, which physically grinds or shears through the material, plasma cutting uses an electrical arc and compressed gas to blast molten metal away, making clean, fast cuts without touching the material (because touching molten metal usually isn't great).

4.1 Why does the industry love CNC plasma?

Well, for one, CNC plasma cutting is fast. Need to slice through 25 mm (1 in) steel plate? A plasma cutter laughs in the face of your cutting challenges. It dramatically reduces production time compared to manual methods like sawing or grinding, and the machine doesn't care if you need one part or a thousand; it cuts them all with the same sharp precision.

Fabrication shops, automotive builders, shipyards, and even metal artists rely on plasma for its incredible speed, repeatability, and sheer cutting power.

Another reason plasma shines (literally) is its cost efficiency. Compared to laser or waterjet cutting, CNC plasma is generally more affordable, both in terms of machine cost and operational expenses.

Consumables are also readily available for CNC plasma cutters and aren't very expensive. Just replace them when they look worn and continue with your daily cutting tasks.

It may not match the pinpoint precision of a laser or handle ultra-thick materials like a waterjet or an oxyacetylene torch, but for most medium to heavy metal fabrication, plasma cutting hits the sweet spot between performance and cost. And let's be honest, showers of sparks make any shop feel a little more tough.

Plasma cutting has a certain primal appeal, and what it makes is very welder-friendly as well, with little additional clean-up.

Once you dial in your settings, your plasma cutter will faithfully turn digital designs into tangible, tough-as-nails parts, art pieces, or structural components, all while you stand back and admire the fireworks.

4.2 CNC Plasma: Perfect with Welding

A CNC plasma machine transforms raw sheets into finished products with electrifying speed and precision.

But the real magic happens when it's teamed up with welding, regardless of the process you'll be using (think GTAW, GMAW, SMAW, etc.).

Plasma handles the slicing, welding seals the deal. Together, they're a powerhouse combo that takes metalwork from concept to completion with clean cuts, tight joints, and bonds that are built to last.

So fire it up, let the sparks fly, and enjoy the controlled chaos that makes this duo indispensable in any fabrication shop, mine included!

4.3 CNC Plasma Safety Myths & Mistakes

Let's not kid ourselves: behind all that flashy power lie plenty of safety myths and mistakes that can turn your shop from a place of creation into a place of chaos (and maybe a few emergency room visits).

So, pull on your welding jacket, crank up the fume extractor, and let's review a few of them.

Myth 1: "It's automated, so I don't need to watch it."

Oh, sure. Just press start and go make yourself a sandwich, right? WRONG. Plasma cutters are not obedient little robots; they are intense, high-temperature arc machines with no sense of humor about bad setups.

Like other CNC machines, they do exactly what you tell them to, even if it's wrong. Extreme heat and warping can make things go even more wrong too!

If the torch snags, the workpiece shifts, or the nozzle wears unexpectedly, you'll come back to a smoldering mess or a torch jammed into the slats. Watch your cuts.

Automation doesn't mean "babysitter-free." It just means you're supervising a machine instead of holding the torch yourself.

Myth 2: "Eye protection? That's for welding, not plasma cutting."

Oh, so you're just going to raw-dog it, staring directly into the bright plasma arc with your naked eyeballs? Yikes.

While plasma cutting doesn't produce the UV intensity of arc welding, it still generates an incredibly bright light, molten sparks, and metal spatter. Without proper shaded eye protection, you can get "arc eye" (a sensation similar to having sandpaper rubbed on your eyeballs) or worse.

I personally wear my full welding gear when CNC plasma cutting, better safe than burned.

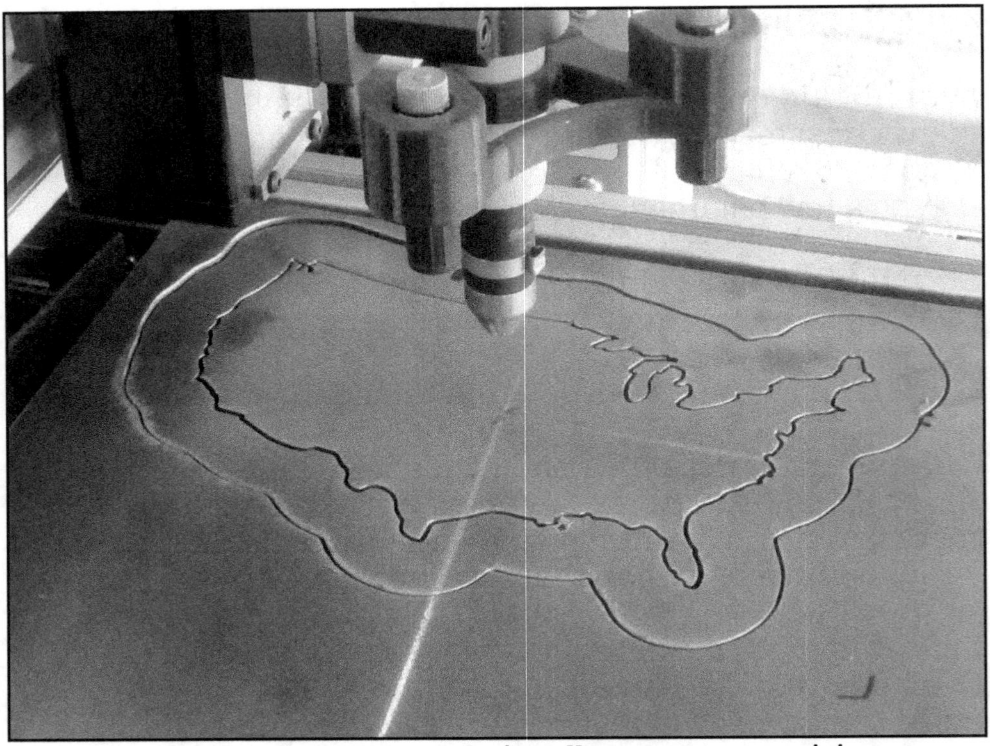

Custom torch LED cross hairs allow greater precision.

Myth 3: "It's fine to cut painted, oily, or rusty metal, the plasma burns it all off."

Technically, yes, the plasma arc will burn through grime, paint, rust, and oily residues. But do you know what else it does? It vaporizes all those coatings into toxic, nasty fumes.

Cutting without proper ventilation or a fume extractor can turn your shop air into a cocktail of heavy metals. Avoid breathing mystery fumes by cleaning your material first and using a good fume extraction system.

Basically, if you smell or see smoke, it will end up in your lungs one way or another.

Myth 4: "I can just hold the piece down with my hand while it cuts."

Look, I know you think you're being clever, but plasma cutting is a high-heat, high-velocity process that can send small parts flying across the table like metal shrapnel.

Holding them down with your bare hand while the arc is engaged is asking to become a viral YouTube injury video.

Use proper clamping, nesting, or tabbing strategies to secure small parts, and keep your hands a very safe distance from the cutting zone.

Myth 5: "The ground clamp can go anywhere."

Like welding, that ground clamp is not a decorative accessory. It's critical to completing the electrical circuit that allows the plasma arc to form properly.

A poor ground connection can cause arc instability, blowouts, erratic cuts, and even torch damage. Always attach the clamp directly to clean metal on the workpiece or table. Clamping it to the rusty paint-peeling corner of the frame is just asking for trouble.

Myth 6: "Shorts and flip-flops are fine, I'm not welding or angle grinding, just cutting."

Sure, wear your best beachwear in the shop, what could go wrong? Oh wait, molten metal sparks, flying slag, and hot debris are raining down like industrial confetti every time you hit start.

Like welding, plasma cutting is not a barefoot-friendly activity. You need long sleeves, closed-toe shoes (preferably leather or steel-toe boots), flame-resistant pants, and gloves. Otherwise, you're just one spark away from involuntary interpretive dance moves.

Myth 7: "Compressed air? As long as it flows, it's fine."

Plasma cutters typically use compressed air as the cutting gas, but here's the kicker: the quality of that air matters.

Moisture, oil, or dirt in your air supply can reduce cut quality, damage consumables, and clog your torch. Think of your plasma cutter like a gourmet chef; it wants clean, dry air so invest in a proper air filter, dryer, and regulator system. This isn't a negotiable.

Myth 8: "I don't need to check consumables; they last a long time."

Plasma consumables (electrodes, nozzles, shields) wear down with every cut, and worn consumables lead to ugly cuts, blowouts, and expensive torch repairs. Ignoring them is like trying to drive a car with bald tires and hoping for the best.

Check your consumables regularly, replace them when worn, and don't cheap out. Running your machine on destroyed tips isn't saving money; it's fast-tracking you to costly downtime. If it seems worn, replace it.

Myth 9: "Grounding and electrical safety? It's just a plasma arc, not high voltage."

Plasma cutters use substantial electrical currents, often in the 40–100 amp range, and poor grounding or faulty wiring can deliver a serious, even lethal, shock to YOU.

Never bypass safety interlocks. Always ensure your machine is properly grounded, and don't mess with the wiring unless you know exactly what you're doing.

The plasma arc may be small, but the power behind it is very real. This is why I just wear my full welding gear when CNC plasma cutting at *CNCROi.com* to keep things easy.

Myth 10: "Maintenance can wait, it's cutting fine today."

The famous last words of many a burned-out torch and seized-up gantry. Plasma systems thrive on regular maintenance: cleaning the table, checking rails and bearings, inspecting torch cables, and keeping the air system in tip-top shape.

Skipping maintenance because "it's working fine now" is like refusing to change your car's oil because "it's running fine today." Sure, you might save time today, but tomorrow you'll be elbow-deep in a repair job you really didn't plan for. Of course, this is especially true when an urgent project is due tomorrow.

One sheet can support many projects, increasing margins.

4.4 Lessons I've Learned About CNC Plasma

A CNC plasma cutter is one of the most powerful and satisfying machines you can have in a shop. But with great firepower comes great responsibility.

Gear up properly, stay alert, respect the consumables, and treat maintenance as part of the craft, not an afterthought.

Remember, you're not just operating a machine; you're commanding a controlled lightning storm to carve metal.

Treat it right, and you'll get beautiful parts, smooth cuts, and a long-lasting machine.

Treat it wrong, and you'll get sparks, smoke, confusion, and possibly a really awkward phone call to the service tech.

So suit up, double-check, and cut smart, because the best plasma operators finish the day with clean parts, an intact machine, and no accidental holes in their shoes.

4.5 Your Plasma Torch Is Having a Meltdown

In the CNC plasma world, consumables are like tires: they wear out slowly, then all at once, usually right when you're cutting your last plate before lunch.

A fresh nozzle gives you razor-sharp cuts. A worn one gives you smoke, slag, and enough dross to build another part, accidentally.

Why Consumables Matter (a Lot More Than You Admit)

That cute little swirl ring and hardened copper nozzle aren't just decorative; they're doing serious work. Together, they focus the plasma, shape the arc, and control airflow, all while operating at around 11,100°C (≈ 20,000°F). When they wear out or get misaligned, even slightly, the entire cut quality goes downhill exponentially.

Worn consumables can lead to all kinds of issues: tapered cuts that don't fit, slag buildup that needs hours of cleanup, and blowouts at pierce points that ruin your material before the cut even starts.

You might notice holes coming out egg-shaped instead of round, and start wondering if your table's out of alignment. But before you blame the machine, take a hard look at the tip as it's often the silent culprit behind the chaos.

The Suspects: What Wears, Why It Matters

Consumable	What It Does	How It Fails
Nozzle (tip)	*Shapes the arc, controls precision*	*Enlarges, pits, melts, arc gets sloppy*
Electrode	*Carries current, forms the arc's heart*	*Gets mushroomed or recessed*
Swirl Ring	*Spins the air/gas for arc control*	*Warps, cracks, or clogs with metal dust*
Shield Cap	*Directs gas, protects arc from backblast*	*Gets spatter-bonded or eroded*

Cut Quality Clues: What Your Torch Is Telling You

If your plasma cuts are starting to look more like abstract art than precision parts, your torch is trying to tell you something, loudly and messily. Most cut quality issues don't come from the table or the software; they come from worn consumables or small setup mistakes that snowball into big problems.

Learning to recognize the signs early can save you time, material, and your sanity.

When the kerf (the width of your cut) starts getting wider, it's a dead giveaway. What used to be a fine, laser-like line now looks more like a trench. This is especially noticeable on thin sheet metal, where once-snug parts now have unintended wiggle room or decorative flair. A widening kerf often means your nozzle is enlarged or damaged and can no longer maintain a tight arc.

Excess dross is another clear sign. If dross used to flake off with just a fingernail and now requires an angle grinder, safety goggles, and a podcast to tolerate, your setup's off. Top-side dross typically points to an issue with gas flow, pierce timing, or cutting height. Bottom dross is more often the result of a worn-out nozzle that can't direct the arc cleanly through the material.

If you notice bevels leaning in one direction or cuts tapering oddly, the arc is likely drifting off-center. This kind of asymmetry is usually caused by a worn or pitted nozzle or a swirl ring that's warped, cracked, or clogged with metal dust. It's not just cosmetic, it throws off dimensions and can ruin fitment, especially on precision parts.

And when your pierces start resembling small explosions rather than clean entry points, it's time to investigate. Blowouts at the start of a cut usually signal a mushroomed electrode or incorrect pierce delay settings. If the arc isn't forming cleanly or the delay is too short, it can blast through the material unevenly, damaging both the part and your consumables.

Torch Autopsy: Inspecting the Victims

When cut quality drops, check your torch first. Worn consumables cause most problems. A quick inspection can save hours.

Nozzle
The orifice must be round. If it's pitted, cracked, or melted, it's done.

Electrode
If the hafnium insert is mushroomed or recessed over 1 mm, replace it or risk damaging the nozzle.

Swirl Ring
Should be clean and intact. Warped, cracked, or burnt? Toss it. Reusing it is like reusing gum, wrong every time.

Cleaning
No sanding, no polishing. Blow out dust, wipe with isopropyl alcohol, and never touch tips with bare fingers.

I use 12 gauge hot rolled mild steel for a lot of my builds.

How Long Do Consumables Last?

Every time your torch fires up and spits plasma, that's a start. It could be a gentle edge start, like easing into a pool from the steps, or a full-on pierce, where the arc has to blast straight through the material like a tiny, molten jackhammer.

Pierces are the heavy lifters here. They're way harder on your electrode and nozzle because you're punching through solid metal, not just gliding in from the side. Think of it this way: edge starts are a polite handshake, pierces are a surprise wrestling match.

In real-world shop life, most starts are pierces, especially in jobs with tight nesting or lots of internal cutouts. And since consumable life is usually rated by how many times the arc is initiated, operators and manufacturers often use "starts" as shorthand. But here's the catch: not all starts are created equal. Every start counts, but every pierce counts double, at least in torch wear and tear.

So next time someone says "you've got 300 starts left," ask: how many of those are just friendly handshakes... and how many are brawls?

Cut Type	Consumable Life	Reality Check
Thin sheet (16–22 ga)	*Nozzle: 300–600 starts* *Electrode: 200–500 starts*	*Depends on pierce delay and air quality*
Mild steel (3 ga)	*Nozzle: 100–200 pierces* *Electrode: 100–200 pierces*	*Life doubles with long, continuous cuts*
Aluminum (4–11 ga)	*Nozzle: 75–150 pierces* *Electrode: 50–100 pierces*	*Aluminum is arc-abusive and clingy*
Stainless	*Nozzle: 60–120 pierces* *Electrode: 50–80 pierces*	*Swirl rings burn out fast under high heat and blowback*

The Smart Operator's Toolkit

Although some people reuse consumables at varying stages of wear, at *CNCROi.com*, I prefer to stay ahead by keeping a full set of unopened replacements on hand. CNC plasma consumables are relatively inexpensive, so it's far better to stay proactive than to gamble on used parts that might deliver inconsistent results.

If you're running production jobs, use a cut log to track when each piece was last changed. It doesn't have to be fancy, even a marker on the side of the torch cart beats guessing.

And whatever you do, don't mix old and new parts. Nozzles and electrodes wear together, like dance partners, if one's tired and the other's fresh, you're asking for skipped steps and ugly cuts. Mismatched consumables don't just affect quality; they shorten each other's lifespan.

Death Comes for All Consumables

Plasma torch parts aren't built to last. They're high-temp, short-lived little warriors that give you clean cuts until they don't. Every arc wears them down, and eventually, they're done. Respect that. Change them before your cuts start looking like fireworks gone wrong.

Like other CNC consumables, these parts don't fail instantly; they fade in cut quality over time. First the kerf widens, then dross piles up, and suddenly you're grinding more than cutting. That's not a machine problem, it's worn consumables asking for mercy.

So when cut quality drops, don't immediately tweak Z-height or gas pressure. Stop. Open the torch. Look at the nozzle and electrode and ask: "Do these look ready to die?"

If the answer's even maybe, swap them.

When it comes to CNC plasma consumables, I just use the brand name ones that come from my torch manufacturer. They cost a bit more, but it's worth the investment.

CNC Machines Have a Body Count

Every CNC machine has consumables, and all of them are slowly wearing out. No matter how well you maintain your setup, something is always inching toward failure, dreaming of retirement, or actively sabotaging your cut.

Got a CNC waterjet? The mixing tube and ruby orifice are basically throwing themselves into a high-pressure sandstorm. They cut steel like champs but wear out like they've seen battle. When your cuts start looking like lasagna, it's time.

Running a CNC router? Your bits are one-job wonders in disguise. They'll swear they've got more life until they snap mid-cut and take your finish (or spindle) down with them.

Using a CNC laser? That soothing hum hides the truth. The tube fades with every pulse, lenses fog like a bad movie, and mirrors slowly collect grime. One day it's crisp cuts, the next, a warped mess of molten acrylic.

And CNC plasma? Your nozzles and electrodes are hurling themselves into high-temperature arcs just to bring you mostly beautiful parts. After a few hundred starts, they'll suddenly ruin your kerf, your fit, and your mood.

Welding? You're not immune. GMAW, GTAW, or SMAW, consumables take daily abuse. Tips wear like drift tires, nozzles clog with spatter, and tungsten gets sharpened more than a prison shiv. Filler rods? Vanish when you need them most. Welding looks permanent, consumables aren't.

Even your CNC 3D printer has traitors. The nozzle's a tiny volcano until it clogs or wears out. The bed goes from sticky to stubborn. The extruder gear gathers grime, and that PTFE tube? Quietly failing with every retraction. Idle printers break too, usually at hour 23 of a 24-hour job.

Whatever machine you're running, something's always dying inside it. Try to spot these early by playing detective and you might just have a lot less headaches while doing projects with your CNC machines in your shop.

4.6 How does arc voltage control improve cut consistency?

Arc Voltage Control (AVC) is basically the torch babysitter in CNC plasma cutting, except instead of wiping noses, it's constantly adjusting the torch height so you don't burn your project (or your reputation). Plasma cutting needs just the right arc length between the nozzle and your metal, think Goldilocks: not too close (or you'll gouge the metal and fry your consumables), and not too far (or the arc just flails around like a party streamer in a wind tunnel).

AVC doesn't sleep. It watches the voltage across the arc like a paranoid hawk because voltage = arc length. If your metal is warped, wavy, lumpy, or just plain defiant, AVC jumps into action like a caffeinated robot gymnast, raising or lowering the torch in real-time to keep the arc happy.

The result? Beautiful, consistent cuts. Less dross (a.k.a. metallic boogers), cleaner edges, longer-lasting consumables, and way fewer swear words from the operator. Without AVC, you'd be eyeballing torch height manually and praying the sheet stays flat, which it never does because metal loves chaos.

4.7 What role does shield gas selection play in cut quality?

Choosing the right shield gas in plasma cutting is like choosing the right fuel for a race car; it dramatically affects how fast you go, how smooth the ride is, and whether your engine explodes halfway through. In this case, though, it's about cut quality, speed, edge smoothness, and whether your consumables age like wine or milk.

Compressed air is the budget-friendly go-to, the instant ramen of plasma gases. It works fine for mild steel and gives decent cuts at a decent speed.

But ask it to handle stainless or aluminum, and it taps out faster than a dad in a TikTok dance challenge. That's because compressed air loves oxidation, and oxidation ruins clean finishes faster than a toddler ruins white pants.

Enter nitrogen or a nitrogen/hydrogen mix, the fancy stuff. This combo is like giving your torch a spa day. It reduces oxidation, improves edge quality, and makes stainless steel cuts smoother than a jazz sax solo. Fancy, right? Now, oxygen is your adrenaline junkie gas. It's hot, it's reactive, and it cuts mild steel like it's slicing warm butter with a lightsaber. The downside? It goes through consumables like a toddler goes through snacks; fast and with zero remorse.

Gas choice also affects kerf width (a.k.a. how wide your cut is), taper (that annoying angle you didn't ask for), and how much post-cut mess you'll be scrubbing off.

Some high-tech systems even let you use mixed gases, like a tag team of elemental ninjas, with one gas doing the cutting and the other shielding the scene like a bodyguard.

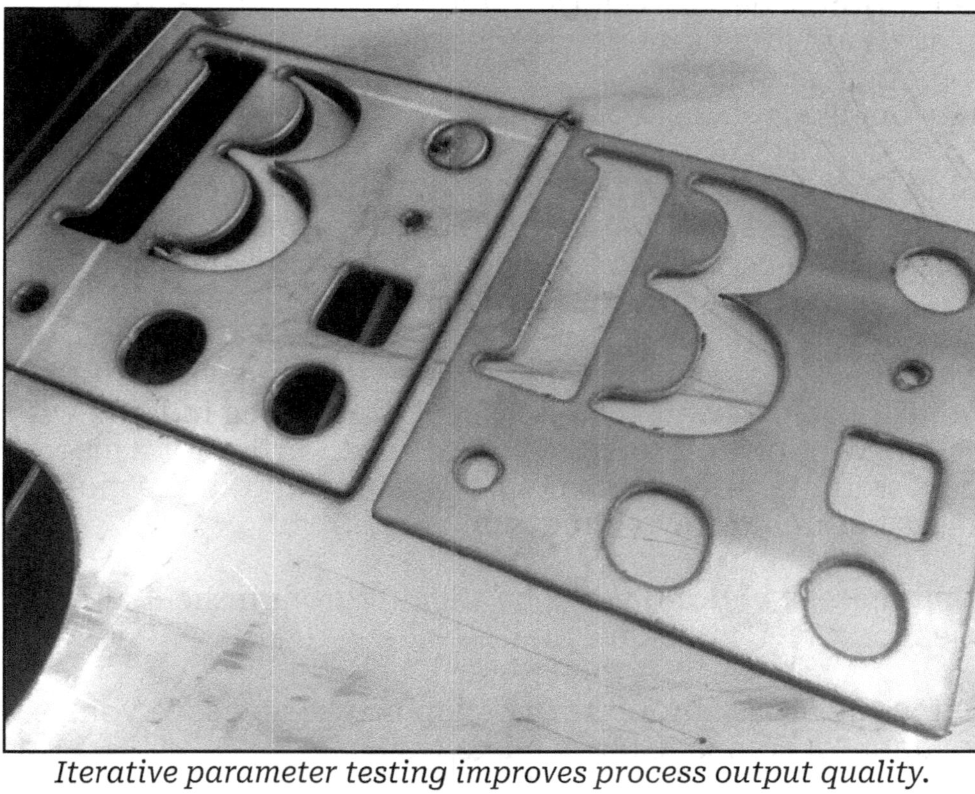

Iterative parameter testing improves process output quality.

4.8 How do you avoid dross buildup on thin sheet metal?

Dross is that gross, crusty blob of molten metal that refuses to let go of the underside of your cut, like the annoying gum on your shoe of fabrication. It's especially annoying on thin sheet metal, where everything heats up way too fast and suddenly your "clean cut" looks like it sneezed out a lava glob.

But fear not, you can avoid most of this metallic menace if you play your settings right.

Step one: Speed is your friend. Cut too slowly, and you're basically slow-cooking your sheet like it's Sunday dinner. All that extra heat invites dross to set up a permanent vacation home under your cut. On thin material, you want to move fast, to stay ahead of the molten chaos.

Step two: Mind the gap. Nozzle height matters. Too far, and your arc's just flailing around like a disco laser show. Too close, and you're melting your way through the material at a sub-optimal height. Get that standoff just right, and your plasma arc hits like a precision-guided lightning bolt, vaporizing metal cleanly instead of spitting it out the bottom like hot slag confetti.

Step three: Chill with the amps. Too much power, and your thin sheet gets absolutely roasted. Use lower amperage and fine-feature consumables to keep things tidy and sharp, like a metal haircut from a surgeon.

Step four: Heat management. Ever try cutting without ventilation or a water table? That's like grilling indoors with all the windows closed. Bad idea. Proper airflow or a nice cool water table helps soak up the heat and trap the molten mess before it throws a tantrum on your cut.

Nail these steps, and you'll be slicing through sheet metal cleaner than a ninja on a diet.

Ditch the dross, impress your shop mates, and save your grinder for someone else's mistakes.

4.9 How do you tune cutting parameters for painted surfaces?

Cutting through paint with a plasma torch is like trying to toast a sandwich that's still in the bag; messy and full of surprises. Paint acts like a heat-blocking force field, messing with your arc and making the torch work overtime.

To cut cleanly:

- Add pierce delay so the arc can bust through that paint armor before moving.
- Crank the amps a bit or slow your roll to keep the cut clean, especially on thick or paint-globbed metal.
- Ventilate like crazy. Burning paint means nasty fumes. PPE isn't optional unless you enjoy breathing chemical soup.
- Expect faster consumable wear, paint gunk = arc drama that's difficult to account for.

Pro tip: If you can, scrape or grind the paint first. A tiny clean strip makes a big difference.

4.10 What's the impact of plasma swirl rings on performance?

Swirl rings may be small, but they're the unsung heroes inside your plasma torch, like nightclub bouncers for the plasma arc. Their job? Spin the gas into a tight tornado around the electrode, keeping the arc focused, centered, and on its best behavior.

A good swirl ring = clean, narrow cuts with less kerf, less taper, and smoother edges. Bonus: your consumables last longer because the arc's not flailing around like it's doing jazz hands.

Different swirl rings suit different jobs, so don't just throw in any ol' donut. And whatever you do, don't ignore them.

A cracked or gunked-up swirl ring leads to bad cuts, unstable arcs, and premature nozzle wear.

Plasma cutting is well suited for producing custom metal tags.

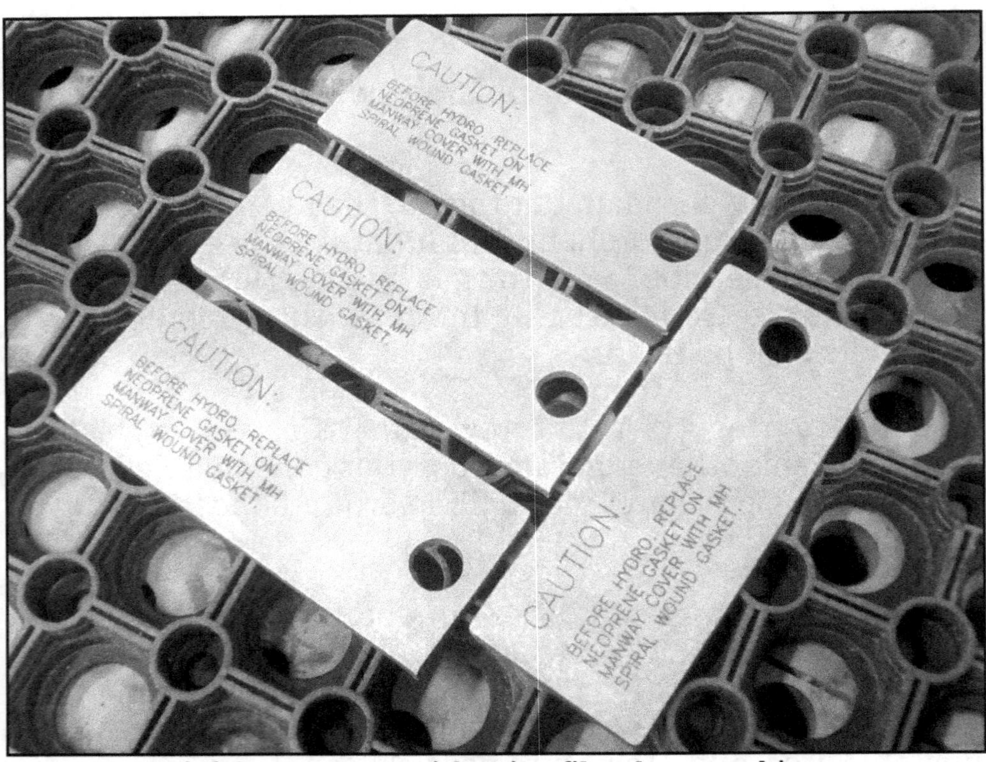

Finish your tags with crisp fiber laser etching.

4.11 How does cutting height calibration work with torch height controllers?

THCs are like the elevator attendants of plasma cutting, always making sure your torch is riding at the perfect floor. Too high? Weak arc and a messy cut. Too low? Nozzle crash and wasted time if the sheet moves.

There are two key heights to nail:

- Pierce height: Higher up, so the torch avoids metal splashback, with perfect aim at YOU!
- Cutting height: Just right for that sweet, focused arc that slices like a hot laser through butter.

Before cutting, the machine politely touches the metal (called initial height sensing) to find the surface. Then it goes to pierce height, fires up the arc, and drops to cutting height to get to work.

The THC uses arc voltage feedback to adjust on the fly, great when your metal is warping from heat or having an uneven surface.

Get the heights wrong? You're looking at dross, torch collisions, and consumables aging faster than a banana in the sun. Get it right? Clean edges, happy torch, longer machine life.

4.12 How do consumable wear rates affect operational costs?

Electrodes, nozzles, swirl rings, they're small, cheap, and absolutely critical. When they wear out, your cuts get ugly, slow, and expensive.

Worn parts = rough edges, fat kerfs, dross everywhere. Ignore them, and you're re-cutting jobs and burning money. Track usage, swap them early, use clean air and the right gas. A good THC setup helps too (no more tip crashes!).

Treat consumables like coffee filters, change them before things get gross.

4.13 What strategies help minimize heat-affected zones?

The Heat-Affected Zone (HAZ) is the cranky neighbor to your plasma cut, not part of the action, but still totally affected. It gets cooked just enough to become harder, more brittle, or warped, especially if you're cutting high-strength steel that doesn't like surprises.

Want less HAZ? Cut fast. Speed = less heat = less problems. Use the right amps, enough to slice, not enough to roast.

Fine-feature consumables and proper gas (like nitrogen or argon) help keep things cool and classy.

Avoid too much piercing (linger too long and you're just flamethrowing the same spot), and cool things down after cutting, air blasts, water baths, interpretive dance, whatever works.

Less HAZ = happier metal, better parts, and fewer surprises when things go under stress.

4.14 How do you handle piercing thick metal plates?

Cutting thick steel (1" and up) with plasma isn't just "point and zap", it's an art form with a side of patience. Rush it, and you'll toast your consumables faster than a bagel on high.

Step one: Use a long pierce delay, thick plates need time for the arc to dig its way through.

Step two: Start with a higher pierce height to dodge backsplash (aka molten metal uppercuts).

Step three: Consider starting with lower amps, then ramp up, kind of like warming up before sprinting with fire.

Pro move: For really thick stuff, skip the plasma pierce entirely and pre-drill a starter hole. Your torch will thank you. Also, ditch shop air, use oxygen or nitrogen for a cleaner, deeper cut.

4.15 How do different electrode materials affect plasma cuts?

Electrodes are the plasma cutter's spark plug, small, hot, and way more important than they look. Most are copper with a hafnium or tungsten insert, because plain copper would melt faster than cheese on a grill.

Hafnium is the MVP, it forms a tiny pit where the arc lives. But as it wears out, that arc gets wobbly, cuts get sloppy, and kerfs go wide. Replace it before your machine starts cutting like it's drunk.

Specialty electrodes such as zirconium or silver provide longer life or improved performance with specific gases or amperage ranges. High-def systems? They're picky, they want matched sets for those super-sharp, precise cuts.

Use the right electrode, or you'll waste time, ruin cuts, and burn through money. Avoid low-quality electrodes; poor materials can reduce performance and consistency.

Consistent circular accuracy indicates correct machine calibration.

4.16 How do plasma tables handle thin sheet fixturing?

Cutting 18–22 gauge sheet metal on a plasma table is like trying to slice a pancake midair with a lightsaber, it can be done, but you're going to need some tricks.

First up: fixturing. Without proper support, thin sheets sag like overcooked spaghetti. So most tables use metal grates or slats to cradle the sheet during the cutting process. Add some edge clamps or magnets to keep things in place if needed, especially if the sheet's got a mind of its own.

Then comes the hero of the story: the water table. This unsung metal-slicing spa bathes your sheet in coolness, absorbing heat, reducing warping, and silencing annoying vibrations. If water's not your thing, a downdraft table pulls air downward like a gentle vacuum, think of it as metal yoga with a fan on. This is the method I prefer to use at *CNCROi.com* for various reasons already outlined.

And don't forget to tune your settings like a plasma DJ: lower the amps, crank up the speed, and use fine-feature consumables. It's like cutting with a scalpel instead of a chainsaw, cleaner, cooler, and way more impressive.

Do all that, and with a little practice, your plasma table will slice through thin sheets so smoothly it'll look like it's showing off.

4.17 What's the advantage of fine-feature consumables?

Regular plasma consumables are like shotguns, great for blasting through thick metal, but not exactly known for finesse.

Enter fine-feature consumables: the precision snipers of the plasma world.

These high-tech doodads (nozzles, electrodes, and shields) are engineered to tighten up the plasma arc like a laser-focused death beam. The result? Super clean, narrow kerfs, crisp edges, and minimal dross, like your plasma cutter just graduated from art school.

They're the go-to gear when you're cutting tiny holes, intricate outlines, or anything that would make a CNC operator cry using regular consumables. Think of it as jewelry-making with a lightsaber.

Bonus: less cleanup! With fine-feature consumables, your parts come off the table looking so good you might just skip the grinder and go straight to showing them off at parties. (Okay, metal fabrication parties, but still.)

Of course, these little wonders have limits. Don't try using them to slice through battleship hull plating. They prefer the calm, manageable world of low to moderate amps and thin materials. Push them too hard and they'll wear out faster than a budget drill bit going through a W beam.

Use them right, though, and you'll be pulling off plasma cuts so sharp and accurate, people might accuse you of secretly switching to laser.

4.18 How do you clean plasma-cut edges for welding?

Plasma-cut metal may look cool, but before you weld, it needs cleanup; unless you enjoy porosity, cracking, and rework.

Step one: dross removal. That hardened slag underneath? Knock it off using a chipping hammer, slag scraper, or flap disc. For delicate work, try a deburring tool or carbide scraper.

Next: oxidation and heat tint, especially on stainless or aluminum. Wire brushes (manual or powered) and Scotch-Brite wheels work great. For critical welds, use pickling paste to strip the oxide layer.

Aluminum alert: always use a stainless-steel brush dedicated to aluminum only; no cross-contamination unless you like weld defects.

Finish with a degreasing solvent wipe to remove oils and fingerprints. Proper prep means better welds, stronger joints, and fewer headaches.

4.19 What are the trade-offs between air and nitrogen cutting gases?

Plasma cutting gases have personalities. Air is the scrappy budget hero, cheap, and perfect for cutting mild steel. Just hook up shop air and go. It does oxidize edges a bit and loves to leave dross on stainless or aluminum, but if you're grinding or painting after, who cares? Bonus: it devours consumables like snacks at a welding party.

Nitrogen, on the other hand, cuts stainless and aluminum with smoother edges and less oxidation, and it treats your consumables gently. The catch? You've got to bring in bottles or bulk supply, more cost and more hassle. Also, it's not as fast on mild steel.

In the end, it's all about priorities: air is cheap and easy, while nitrogen is clean and fancy, so choose wisely, or just blame the gas when your edges turn out ugly.

At *CNCROi.com*, I just use compressed air BTW.

Need tiny round holes not possible with plasma? Use a drill press.

4.20 How do you reduce noise levels during plasma cutting?

Plasma cutting is loud. Between the screaming plasma arc, rushing air, and roaring exhaust fans, it's less "zen workshop" and more "metal thunderstorm."

Start with a water table, the unsung hero of plasma peace and quiet. It soaks up both sound and vibration, making your table sound more like an angry faucet than a rocket launch.

Next, enclose the chaos. Sound-dampening curtains or panels won't silence the torch, but they'll keep the noise from terrorizing everyone within a five-shop radius.

Also, keep your fans, compressors, and exhaust systems maintained, because nothing says "professional shop" like a compressor that doesn't scream like it's being chased.

Finally, equip your crew with noise-canceling headsets or solid hearing protection. Plasma might never be whisper-quiet, but with the right setup, it doesn't have to sound like the apocalypse either.

4.21 How can you integrate CNC plasma systems with robotic arms?

Slap a plasma torch on a 6-axis robot, and suddenly you're not just cutting flat sheets, you're slicing pipes, bevels, beams, and bizarre 3D shapes that flatbed tables can only dream of.

Of course, it's not plug-and-play. You'll need fancy CAM software to choreograph the robot's every move, keeping the torch angle, height, and speed just right while dodging collisions with work clamps and weights.

Smart systems even add height sensing to handle warped surfaces, while tight communication with the plasma unit keeps arc timing, gas flow, and pierce delays perfectly synced. The fancier your setup, the more it will cost and the more training it will require. There's no free lunch here.

4.22 What's the benefit of water tables vs. downdraft systems?

Plasma cutting creates smoke, sparks, and chaos, so shops use either water tables or downdraft systems to keep the air (and your sanity) clear.

Water tables are like metal-cutting hot tubs. They catch sparks, cool the part, muffle noise, and trap fumes before they ever hit your nose. They're great for enclosed spaces or when your shop wants that spa-like "low warping, low drama" vibe. Bonus: less heat-affected zone, which means prettier parts.

Downdraft tables, like the ones *CNCROi.com* uses, are the vacuum cleaners of the plasma world. They suck fumes straight down through the table into an exhaust system, keeping the shop dry and your lungs happy, with no water, no sludge, and no mystery rust puddles.

Want less noise and heat? Go with water. Prefer dry feet and clean filters? Downdraft's your buddy. Either way, your plasma cutter gets to party, and you get to breathe easy. Win-win.

4.23 How does amperage setting influence cut taper?

Too little amperage, and your plasma arc cuts like a dull chisel, leaving tapered edges and wobbly walls.

Too much, and sure, it punches through like a beast, but now you've got wide kerfs, extra dross, and consumables burning away at an exponential rate.

Get it just right, and you've got a tight, focused arc slicing clean, straight walls with minimal taper.

Of course, amperage doesn't work alone. Speed, standoff, and gas flow, along with gas type, all join the party.

Nail the settings, and you'll spend less time grinding and more time admiring your perfectly cut parts.

4.24 What machine upgrades extend consumable lifespan?

Consumables aren't cheap, so treat them right and they'll stop ghosting you mid-cut like a bad job interview.

Start with torch height control. It's like cruise control for standoff distance, preventing nozzle faceplants and uneven wear. Fine-feature consumables also help by focusing energy where it counts, not all over the place.

Next, feed your torch clean air. Gas filters and air dryers remove the gunk (oil, water, mystery particles) that chew through electrodes like candy. Add water mist systems for bonus cooling and longer life.

Smart CAM software can optimize pierces and paths so you're not wasting arc time like a runaway grinder. And don't forget maintenance tools, since alignment jigs and wear gauges help you swap parts when it's actually time, not just when someone "has a feeling."

4.25 How do CNC plasma and oxy-fuel systems complement each other?

Plasma and oxy-fuel are like the dynamic duo of metal cutting, each with its own strengths, and together, unstoppable.

Plasma is your fast, flashy precision cutter. It zips through thin to mid-thick steel, stainless, and aluminum with laser-like finesse (well, plasma-like). Clean edges, tight curves, and non-ferrous metal? Plasma's your pal.

Oxy-fuel is the slow but strong heavyweight champ. It eats thick mild steel for breakfast, several inches thick, with low operating costs and no need for a massive power hookup. Just don't ask it to touch stainless or aluminum. That's not its thing.

Put them together on one CNC table or in the same shop, and you've got a dream team: plasma for speed and detail, oxy-fuel for brute force and budget cuts.

4.26 How do you fine-tune pierce delay settings?

Pierce delay is that crucial pause where the torch says, "Hang on, let me melt through this first," before moving. Get it wrong, and things go sideways.

Too short?
The torch jumps the gun, causing incomplete cuts, ragged starts, and confused operators.

Too long?
It melts a crater to the Earth's core, throws dross everywhere, and cooks your consumables.

The fix?
Start with the cut chart, adjust in small steps, and watch that pierce point like it owes you money.

Smart CAM systems can even auto-adjust delays based on pierce location, because piercing near an edge is not the same as blasting through plate armor in the middle.

Bow tie inlays made from 2B 304 stainless with router a jig.

4.27 What are the pros and cons of mechanized vs. handheld torches?

In the world of metal cutting, mechanized plasma torches are basically the overachieving, straight-A students of the shop floor. They ride CNC machines like futuristic chariots, slicing through metal with laser focus (well, plasma focus, but you get it).

These torches are precise, consistent, and unfazed by complexity. With features like height control, gas regulators, and consumables fancier than a gourmet chef's knives, they're made for high-speed, high-stakes production.

Then we've got the handheld plasma torches, the rebellious artists of the cutting world. They do not follow programmed paths; they rely on instinct rather than G-code. Want to cut a rusty tractor part in a barn at midnight? No problem. The handheld torch has your back. They're portable, affordable, and don't require a PhD in CNC programming. Just plug them in, PPE up, and start slicing like a hot knife through steel.

Mechanized Pros: Super accurate, repeatable, and fast, if you like your cuts like you like your teas: consistent.

Mechanized Cons: Pricey, complicated, and they don't laugh at your jokes.

Handheld Pros: Cheaper, flexible, and great for on-the-fly fixes.

Handheld Cons: Your arm is the automation. Also, ut quality depends on you so you have no excuses.

In most shops, it's not an either-or situation. It's like having a Ford Lightning and a dirt bike. One's for the freeway, the other's for when you want to jump a ditch.

Together, they make your operation sharp, smart, and ready for anything, whether it's a precision job or an impromptu repair mission in the rain. I use both at *CNCROi.com* on a very regular basis, both have their place in my heart.

4.28 How do you calculate part shrinkage due to thermal input?

Plasma cutting gets metal really hot. So hot, in fact, the metal expands mid-cut like it's hitting the sauna. Then it cools down and shrinks, just a little, but enough to mess with your "perfect" dimensions if you're not careful.

This shrinkage is usually small, but in high-precision or multi-part jobs, it can turn "nice fit" into "why doesn't this bolt line up?"

If you're a math nerd, you can use the thermal expansion coefficient (like 12×10^{-6} per °C for mild steel), multiply it by your part length and temp change, and boom, expansion math.

If you're a shop-floor realist, you just cut the thing, measure it, and tweak the program until it works. Both methods are valid. The key takeaway: metal moves when it's hot, so plan accordingly unless you like playing "guess that gap" during assembly.

4.29 How does consumable cooling extend electrode life?

Plasma cutting is hot. Like, melt-your-face hot. The poor electrode sits in the middle of it all, enduring a hellish combo of heat and electricity. Without good cooling? It burns out faster than a cheap sparkler on the 4th of July.

That's why plasma torches use liquid, gas, or air cooling, to keep things chill and prevent the electrode (especially the fragile hafnium bit) from frying. If cooling fails, your arc gets jumpy, cuts get ugly, and your consumables tap out early.

Pro tip: To keep your plasma torch running cool and your consumables lasting longer, always check that coolant is flowing properly, clean out any gunk or buildup in the cooling passages, use clean, filtered gas to avoid contamination, and whatever you do, don't cheap out on the cooling system. A little maintenance goes a long way.

4.30 What methods improve cut quality on rusty or painted surfaces?

Cutting through rusty or painted metal is like trying to sprint in flip-flops, messy, unpredictable, and likely to trip you up. Surface gunk like rust and paint interferes with arc stability, causing extra dross, rough edges, or cuts that just stop halfway. Best fix? Hit the area with a wire wheel, grinder, or scraper. A little cleanup goes a long way.

If cleaning's not an option, tweak your settings. Bump the amps to punch through the grime, extend the pierce delay so the arc can establish, and slow your travel speed to keep things stable.

Bonus tip: High-end systems with pilot arc or high-frequency starts handle dirty, nonconductive surfaces better. And don't forget the work clamp, as a bad ground means bad cuts. Sure, plasma can cut through crusty junk, but clean metal means better cuts, longer consumable life, and fewer shop-floor frustrations.

Flat plates prepared for secondary fiber laser processing.

4.31 How do you manage arc start failures?

Arc won't start? Welcome to plasma's version of a bad Monday. Here's how to kick it back into gear without losing your mind (or production time):

First off, check that ground clamp, it needs to bite clean, bare metal. No rust, no paint, no excuses. Next, peek at your consumables. If your electrode or nozzle looks like it had a rough night, swap it out. No spark, no cut.

Air quality matters too. If your plasma torch is huffing dirty, wet air, it's not gonna perform. Use dry, filtered air, not swamp gas. Also, double-check the starter system: high-frequency, contact start, whatever your setup is, make sure it's not fried.

Oh, and make sure you're not trying to cut glass. Plasma needs something conductive, not your lunch tray or that mystery-coated sheet metal.

Stay sharp, stay grounded, and keep your torch drama-free.

4.32 How do high-frequency start vs. contact start systems differ?

High-frequency (HF) start and contact start are like the two weird cousins of plasma arc ignition, both get the job done, but in very different ways.

HF start fires off a high-voltage spark to jump-start the arc without touching the workpiece. Great for cutting everything from steel to sci-fi alloys, but it throws off enough electromagnetic interference (EMI) to make your Wi-Fi cry. It's old-school, industrial, and not super friendly around sensitive electronics.

Contact start, on the other hand, is more polite. It taps the electrode and nozzle together inside the torch like a little plasma handshake, then pulls apart to light the arc. Cleaner, quieter, and way better for CNC setups or anything near electronics that shouldn't glitch out.

4.33 What role do height sensors play in preventing tip crashes?

Height sensors in CNC plasma cutting are basically the torch's sixth sense, keeping it from face-planting into warped metal like a drunk Roomba.

These little lifesavers use capacitive, ohmic, or arc voltage feedback to constantly check the material's height and tweak the torch's Z-axis on the fly. Before the first spark, they find the right pierce height; during the cut, they play bouncer, keeping the torch floating perfectly over lumps, bumps, and metal doing the heat-warp shuffle.

No sensor? Say hello to tip crashes, fried consumables, busted torches, and some very expensive scrap art.

With a good torch height controller (THC) and a sensor that's paying attention, you get clean cuts, longer part life, and fewer repair bills. It's like autopilot for your plasma, minus the crash landings.

4.34 How can you design parts to minimize cut time on thick plates?

First rule: pierce less, stress less. Every pierce eats time, wears consumables, and invites trouble. Combine features or use shared edges to keep pierces to a minimum, your torch and wallet will thank you.

Next up: keep it smooth. Plasma can do fancy curves and tight corners, but every sharp detail makes it pause and think. Smooth, flowing contours = faster cuts and happier machines.

Common-line cutting is your best friend. Let adjacent parts share a cut line, saves time, saves metal, saves your sanity. This is tricky though to get right 100% of the time. And when nesting parts, pack them tighter than a family in a minivan on vacation, fewer travel moves, less wasted space.

Finally, thick plate = slow speeds. So avoid overly long or dainty features that push your machine's patience

4.35 What upgrades improve edge angularity on precision plasma systems?

Edge angularity: the technical term for "is this edge straight, or did someone cut it with a spoon?"

If your parts are looking like they were sliced with a butter knife at a weird angle, it's time to upgrade. First, ditch the basic plasma setup and go high-def, it's like going from dial-up to fiber internet.

High-definition systems shoot a tighter arc, which means less taper and cleaner, crisper edges. Throw in some fine-feature consumables, and suddenly your plasma torch has the precision of a neurosurgeon on Ritalin, steady hands, sharp focus, and zero tolerance for sloppy cuts.

Then there's torch height control (THC), think of it as cruise control for your torch. With arc voltage feedback, it keeps the standoff distance in the sweet spot, so your arc stays focused and doesn't wander off like a distracted toddler. Combine that with tight motion control, precision rails, ball screws, or servo motors, and now your cuts look engineered, not guessed. No more plasma Picasso edges or customers wondering if you used a hand torch isntead.

Gas matters too. The right cutting gas (or mix) can make all the difference, especially when cutting aluminum or stainless. Think of it like seasoning a steak, bad gas = bad flavor (or in this case, bad edges). Use the good stuff and your arc stays sharp and consistent, instead of flaring out like a birthday candle in a wind tunnel.

Finally, let your CAM software do the thinking. The right program will dial in your speed, amperage, and gas flow like a plasma-cutting DJ spinning all the right settings. Bonus points if it can auto-compensate for plate thickness or suggest the best lead-in paths.

Put it all together and you're not just cutting metal, you're crafting precision parts that make laser cutters side-eye you in envy while whispering, "Okay, plasma... I see you."

4.36 CNC Plasma: Top 10 Takeaways

Arc voltage control improves cuts: *Automatic voltage regulation maintains consistent cut height, reducing taper and improving edge quality on warped or uneven sheets.*

Shield gas affects cut quality: *Different gases like air, nitrogen, or oxygen impact edge finish, cut speed, and dross levels, choose based on material and application.*

Dross buildup is a common issue: *Fine-tune cutting speed, torch height, and gas type to minimize dross for faster cleanup, better weld prep, and less rework.*

Piercing thick metal needs care: *Controlled pierce delays and power settings help prevent nozzle blowback, protect consumables, and ensure a clean start.*

Fine-feature consumables improve detail: *Specialized tips and electrodes maintain arc stability on small radii, enabling clean cuts on intricate shapes and tight contours.*

Water tables vs. downdraft: *Water tables suppress noise, cool parts, and trap particles; downdraft systems actively extract fumes and reduce airborne dust in dry setups.*

Consumable wear affects cost: *Worn or damaged tips degrade quality, increase scrap, and slow production, inspect regularly and replace parts before failure.*

Heat-affected zone management: *Proper control of speed, amperage, and cooling minimizes warping, discoloration, and structural changes in sensitive materials.*

Robotic integration increases precision: *Automated arms and motion systems boost accuracy, improve cycle times, and enhance repeatability for high-volume production.*

Height sensors prevent crashes: *Sensors constantly adjust torch height to maintain proper standoff, avoiding tip collisions, damage, and costly downtime.*

✖ 4.37 Section Quiz

This quiz is like a CNC plasma cutter, take the wrong path, and you're headed straight for a tip crash and a pile of scrap!

1. What does arc voltage control help regulate in plasma cutting?
    A) Shield gas composition
    B) Torch height during cutting
    C) Abrasive flow
    D) Cooling water temperature

2. Why is shield gas selection important?
    A) It determines table size
    B) It reduces motor vibration
    C) It affects cut quality and edge smoothness
    D) It increases software compatibility

3. What is dross?
    A) A type of shielding gas
    B) The molten residue left on the underside of the cut
    C) A software setting
    D) The cooling water used in the torch

4. How can you minimize dross on thin sheet metal?
    A) Increase amperage and slow down
    B) Decrease cutting speed
    C) Optimize speed and torch height
    D) Use higher-pressure water

5. What is the function of a swirl ring in a plasma torch?
    A) Directs plasma gas flow for a tighter arc
    B) Holds the electrode in place
    C) Cools the nozzle
    D) Stores shield gas

6. Why is proper cutting height calibration important?
    A) It reduces table noise
    B) It changes plasma gas composition
    C) It improves software updates
    D) It maximizes consumable life and cut precision

7. How do consumable wear rates affect operational costs?
   A) Faster wear increases long-term costs
   B) Slower wear raises gas costs
   C) Wear rates have no cost impact
   D) Wear only affects software updates

8. What helps minimize the heat-affected zone (HAZ)?
   A) Using lower-amperage settings and higher speed
   B) Increasing arc voltage
   C) Cutting at slow speeds
   D) Increasing standoff distance

9. How do you pierce thick metal plates effectively?
   A) Use the same settings as thin metal
   B) Reduce amperage
   C) Use a slow pierce with delayed motion
   D) Start cutting from the edge only

10. What are fine-feature consumables designed for?
   A) Higher speed only
   B) Maximizing rough cuts
   C) Cutting small, detailed parts with better accuracy
   D) Reducing overall power usage

11. How should you clean plasma-cut edges before welding?
   A) No cleaning is needed
   B) Sandblast the part
   C) Soak in water
   D) Wire brush or grind off oxide layers and dross

12. What's the difference between air and nitrogen cutting gases?
   A) Nitrogen cuts slower but makes cleaner edges
   B) Air is faster on nonferrous metals
   C) Air produces no heat
   D) Nitrogen reduces consumable life

13. What's the benefit of using a water table under a plasma machine?
   A) It reduces noise, fume, and heat distortion
   B) It increases cutting voltage
   C) It sharpens the electrode
   D) It raises arc temperature

14. How does amperage setting influence cut taper?
   A) Higher AMPS widens the top and narrows the bottom
   B) Higher amperage reduces taper by fully penetrating
   C) Lower amperage increases cut speed
   D) Amperage has no effect on taper

15. How can you extend consumable lifespan?
   A) Increase amperage beyond rated limits
   B) Ignore torch height settings
   C) Use correct cutting parameters and gas flow
   D) Reduce cooling air

16. How do CNC plasma and oxy-fuel systems complement each other?
   A) Plasma cuts thick; oxy-fuel handles thin sections
   B) Plasma replaces oxy-fuel completely
   C) Both systems are identical
   D) Oxy-fuel cuts thick steel; plasma handles thinner

17. What is pierce delay?
   A) A safety feature
   B) A setting that penetrates before motion begins
   C) A type of consumable
   D) A maintenance procedure

18. What's one benefit of mechanized over handheld torches?
   A) Reduced table weight
   B) Higher cutting power
   C) More precise, repeatable cuts
   D) Lower software requirements

19. Why are height sensors important?
   A) To prevent tip crashes and keep consistent cut height
   B) To control shield gas temperature
   C) To adjust cutting gas composition
   D) To monitor operator fatigue

20. How can you reduce cut time on thick plates?
   A) Increase torch height
   B) Use slower speeds
   C) Optimize part design and toolpaths
   D) Reduce amperage

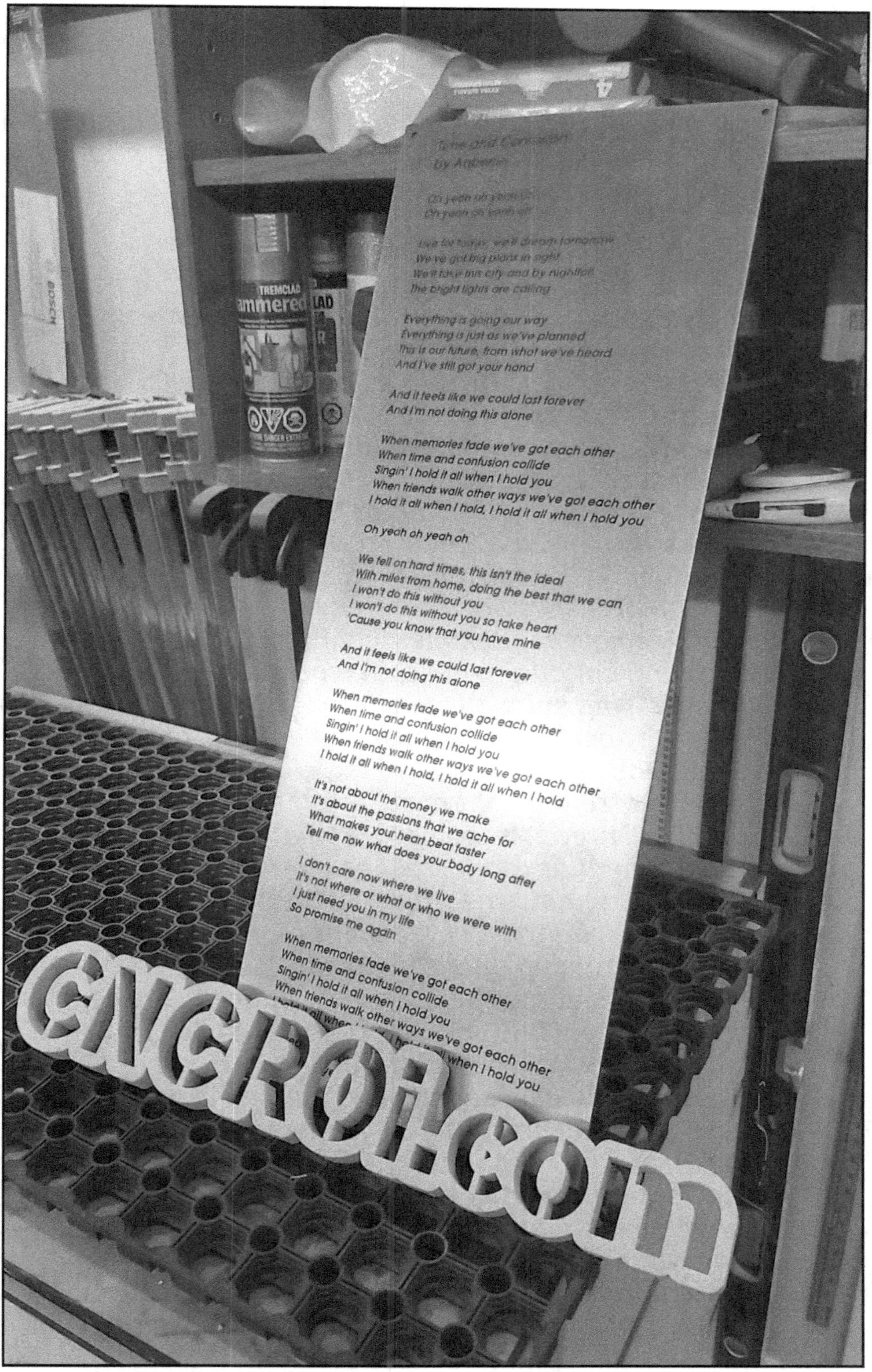

Lyrics sheet cut on the plasma and etched with a fiber laser.

✖ 4.38 Answer Key and Explanations

This answer key hits harder than a tip crash on a warped plate, but don't worry, the explanations cut cleaner than a dialed-in torch on fresh steel.

1B – Arc voltage control adjusts torch height automatically to maintain optimal standoff, ensuring consistent cuts, reducing taper, and preventing tip crashes on uneven material.

2C – Shield gas impacts arc behavior, dross levels, and edge finish. Air is cheap but messy; nitrogen offers cleaner cuts on stainless or aluminum; oxygen works well on carbon steel.

3B – Dross is leftover molten metal that sticks to the part. It's usually caused by incorrect torch height or speed and adds time for cleanup.

4C – Proper travel speed and torch height allow the arc to cleanly remove molten metal, minimizing dross and improving cut quality on thin materials.

5A – The swirl ring spins the gas into a vortex, focusing the arc for better precision, cleaner cuts, and reduced arc instability.

6D – Correct cutting height ensures arc focus, reduces taper, and prevents premature wear from arc misalignment or torch crashes.

7A – Worn consumables degrade cut quality, increase scrap, and lead to more frequent replacements and downtime, raising operating costs.

8A – Less heat input and faster travel reduce the heat-affected zone (HAZ), preventing warping or metallurgical issues near the cut.

9C – Thick metal needs a controlled pierce delay to avoid blowback and ensure the arc penetrates fully before cutting begins.

10C – Fine-feature consumables produce a tighter arc for cleaner, more accurate cuts on intricate parts and tight curves.

11D – Cleaning removes slag and oxides, improving weld quality and preventing contamination or weak welds.

12A – Nitrogen creates smooth, low-oxidation edges on stainless; air is faster but messier, especially on reactive metals.

13A – Water tables absorb heat, trap fumes, and reduce noise, improving safety and part quality, especially on thin materials.

14B – A strong arc cuts straight through the material, creating cleaner, more vertical edges with less taper.

15C – Following recommended settings extends consumable life by maintaining arc focus and reducing unnecessary wear.

16D – Plasma and oxy-fuel each excel in different applications, offering flexibility across material types and thicknesses.

17B – Pierce delay ensures the arc starts cleanly by allowing it to fully pierce the material before torch movement begins.

18C – Mechanized torches offer consistent motion and height control, producing more accurate, repeatable parts than handheld cutting.

19A – Height sensors adjust the torch automatically during cuts, avoiding collisions and ensuring stable arc length.

20C – Efficient nesting, fewer pierces, and shared cut lines reduce travel time, cut time, and consumable wear on thick plates.

My custom tea light business was big business at craft shows.

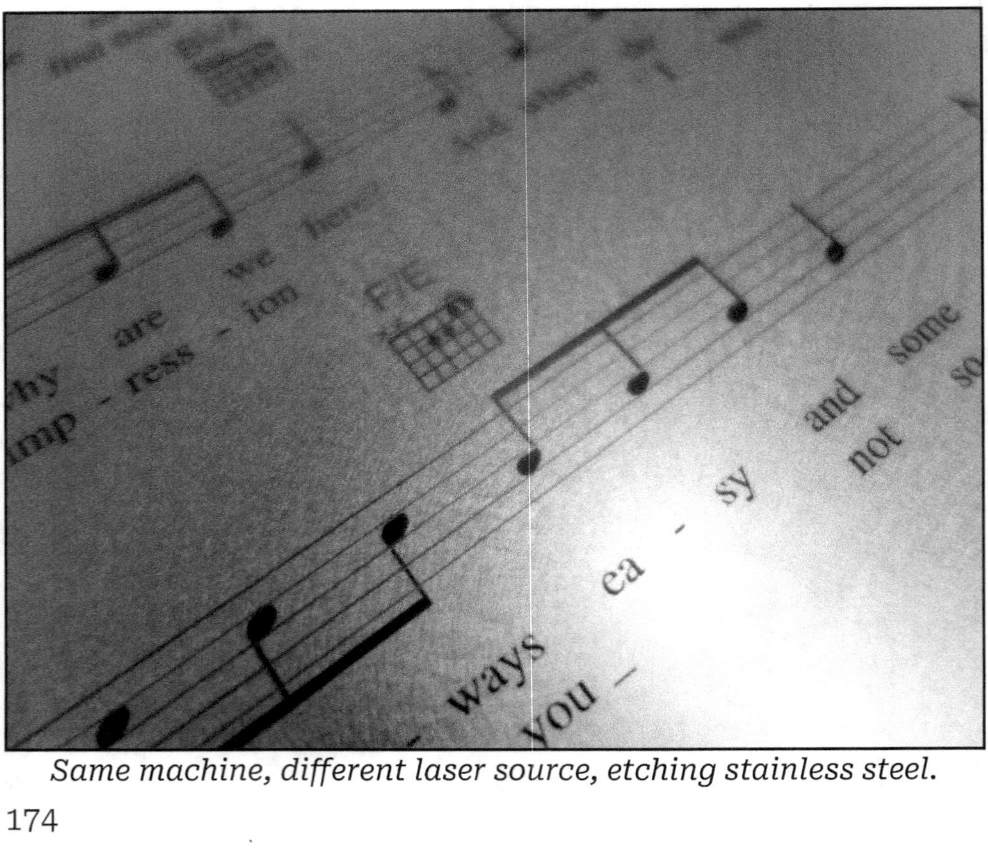

Same machine, different laser source, etching stainless steel.

Section 5
The Laser Bay
Dialing In Passes for Clean Cuts

CNC laser cutting is basically what happens when you strap a lightsaber to a robot and tell it to carve out your wildest designs with surgical precision. It's sleek, it's high-tech, and it's probably the coolest way to cut materials without getting your hands dirty (or accidentally setting something on fire, assuming you use it correctly).

Whether you're etching tiny details onto a piece of wood or slicing through sheet metal with blinding speed, CNC lasers have become a go-to tool for industries that demand clean, precise, and repeatable results.

CNC laser cutting works by focusing a high-powered beam of light onto a material, using heat to vaporize, melt, or burn away what's in the path.

The CNC controller moves the laser head along X and Y axes (sometimes Z, too), following the digital design file with a level of accuracy human hands could never match.

5.1 Why does industry love lasers?

Lasers are precise, we're talking tolerances down to fractions of a millimeter. That makes them ideal for industries like aerospace, electronics, medical devices, and automotive manufacturing, where even the tiniest imperfection can spell disaster.

Unlike plasma, which is great for thicker materials but rougher cuts, or waterjets, which are slower but cut anything, CNC lasers excel at crisp, clean edges on thin to medium sheet materials.

Another huge perk is the versatility. A CNC laser doesn't just cut, it can engrave, mark, and etch depending on your settings.

That means you can personalize products, label parts, or add decorative details without changing tools or setups.

Want to cut a stainless steel part, then mark it with a barcode or serial number? No problem. Want to engrave intricate artwork onto wood, leather, or acrylic? Easy.

Basically, lasers are like the multitool of the CNC world, but way more precise (and let's face it, way cooler).

Of course, one of the best parts of working with a CNC laser is the sheer fun of it. Watching that little invisible beam glide across the material, effortlessly cutting or engraving, feels like science fiction brought to life.

There's no physical contact, no tool wear, and no sparks flying, just the quiet hum of precision technology doing its thing while you sip your coffee and admire how futuristic your shop suddenly feels.

5.2 Versatile Beams, Limitless Applications

CNC laser cutting is a cornerstone of modern manufacturing, blending speed, precision, and versatility in machines that wouldn't look out of place on a sci-fi set.

From high-powered fiber lasers slicing through steel and aluminum, to versatile CO_2 lasers etching wood, acrylic, and leather, to ultra-precise UV lasers marking glass, and plastics, there's a specialized laser for almost every material and use case. Whether you're crafting rugged industrial components, delicate jewelry, intricate signage, or rapid prototypes, CNC lasers turn raw stock into refined, ready-to-use products with astonishing accuracy.

What sets these systems apart is their adaptability. With the right settings and optics, a single machine can jump between tasks with minimal downtime, making them invaluable in both high-volume production and creative one-off builds. So go ahead, load up your design, calibrate your beam of choice, and enjoy the glow of one of the most powerful, flexible technologies in the CNC universe.

5.3 CNC Laser Safety Myths & Mistakes

CNC laser, the sleek, futuristic machine that slices, engraves, and vaporizes materials with light.

It's the closest most of us will ever get to wielding a lightsaber, and let's be honest, it's incredibly satisfying to watch it carve delicate patterns into wood, acrylic, leather, or even thin metal.

Here come 10 safety myths and mistakes about CNC lasers, served with a hearty side of humor.

Myth 1: "It's just a little laser, how dangerous can it be?"

CNC lasers can run from 40 watts (enough to cut through wood and plastic) to multiple kilowatts ranges (capable of cutting metal plate).

That beam can blind you, burn you, and ignite things you didn't even know were flammable. Treat it with respect, this is precision-engineered industrial light, not a glow stick at a rave.

Keep in mind that behind that beam is a machine that can pinch, crush, or injure you just as easily. That little beam is only one small part of a much larger system you must stay aware of and respect.

Myth 2: "I don't need eye protection if I don't stare directly at the beam."

Sure, and you don't need sunscreen if you stay near the Sun, right? Just be careful: even reflected or scattered laser light can damage your eyes, especially when cutting reflective materials like metal or mirrored acrylic. It's like getting a sunburn while lounging in the shade at Cape Tribulation in Queensland (ask me how I know).

Laser goggles or shields (green for fiber) aren't optional fashion accessories; they're essential PPE. You might think you look cool peeking over the machine bed, but you won't look so cool with permanently scorched retinas.

Myth 3: "It's an enclosed machine, so fume extraction isn't really necessary."

Vaporized wood, plastics, leather, and adhesives all release a wonderful bouquet of toxic fumes: formaldehyde, benzene, chlorine, and a host of other compounds you do not want in your lungs.

Even if the machine has a lid, without proper fume extraction or filtration, you're basically turning your workshop into a gas chamber of doom. Vent it outside or install a good fume filter, your lungs will thank you.

Myth 4: "Acrylic smells kinda sweet when it burns, it's probably safe."

Just because it smells like cotton candy doesn't mean it's harmless. Burning acrylic releases methyl methacrylate, which can irritate your eyes, skin, and respiratory system. Fun fact: "sweet smell" is often a warning sign in industrial safety.

Don't sniff the fumes. Don't work nose-deep in the cutting chamber if you want healthy lungs.

Myth 5: "I can cut anything that fits under the lid."

No, you can't just throw any material in there. Just because it physically fits doesn't mean it's safe to laser. For example, PVC releases chlorine gas when lasered, which can corrode your machine, your shop, your lungs, and just about anything else it touches.

Some plastics release cyanide gas. Certain foams ignite almost instantly. Even seemingly harmless materials can produce toxic fumes or leave behind stubborn, corrosive residues.

Always, always check the material compatibility list before you load up a sheet and hit "start." Don't assume, verify.

There are ample websites out there you can input your material MSDS in and get a scare, use them for knowledge.

Myth 6: "If the cut isn't going through, I'll just slow it down and crank up the power."

Sometimes poor cuts aren't about power; they're about dirty lenses, bad focus, or the wrong settings.

Blasting through by upping power can damage the optics, overheat the material, or ignite flame-ups inside the bed. Take a breath, troubleshoot properly, and remember: brute force isn't always the answer.

Myth 7: "I don't need to clean the lenses often."

Dirty lenses act like little laser death traps, scattering and absorbing energy where they shouldn't. This leads to reduced power, bad cuts, and eventually, cracked lenses.

Regular lens maintenance isn't optional, it's the key to keeping your machine sharp, your cuts clean, and your wallet happy. Plus, let's be honest, if you're willing to detail your car but not clean your laser, we need to have a talk.

If it fits, it's a business!

Myth 8: "Laser fires? The beam's too narrow to start a fire."

Yes, lasers cut by vaporizing material, and vaporizing means heat. Flammable materials like paper, thin woods, or certain fabrics can catch fire in seconds, especially if you set slow speeds or run multiple passes.

Always keep a fire extinguisher nearby, and never leave the machine running unattended. Learn from three of my friends who burned down their shop doing just that.

Myth 9: "My workshop's small, so grounding and electrical safety aren't a big deal."

CNC lasers draw a surprising amount of power, and skipping proper grounding or using cheap extension cords is a recipe for tripped breakers, blown boards, or even electrical fires.

Keep your machine on its own grounded circuit, use proper surge protection, and follow the manufacturer's electrical guidelines, follow these like gospel if you want a warranty!

After all, you're playing with focused light beams and high voltage, what could possibly go wrong?

Myth 10: "The software is smart, I don't need to watch the job once it starts."

Software is smart... until it isn't. A misaligned file, a double-layered line, a wrong origin point, any of these can cause your laser to go off-course, cut where it shouldn't, or even slam the head into the frame.

Always preview your toolpaths, double-check your focus and origin, and most importantly, stay nearby while the machine runs.

Automation doesn't mean autopilot, it just means you're supervising a very precise robot. A CNC laser is no different from a CNC router, waterjet, or plasma cutter: they all follow instructions exactly, including the bad ones. If you feed it the wrong settings, it'll still execute them flawlessly.

Outdoor craft shows can grind dirt into your tablecloth.

5.4 Lessons I've Learned About CNC Laser

A CNC laser is one of the most diversified tools in the digital fabrication arsenal. It can cut, engrave, etch, and mark with unbelievable precision, making projects come alive like magic.

Like any powerful tool, it demands respect, attention, and good safety habits.

Skip the goggles, ignore the fumes, leave it running alone, or go "mystery material" mode, and you're not just risking a ruined project, you're risking injury, machine damage, or even a shop fire.

But use it wisely, follow the rules, and embrace the (careful) magic, and you'll have a machine that pays you back with crisp lines, detailed cuts, and a long, happy life.

So, clean those lenses, vent those fumes, and channel your inner laser Jedi, safely. May the (laser) force be with you!

5.5 A Dirty Lens Wrecks Everything

Lasers are magical, focused light, cutting through metal, wood, and just about anything else depending on the source type. But even the most powerful laser turns into a flashlight if your lens is filthy, cracked, or installed backwards (yes, it happens).

The Lens is the Real MVP (Most Vulnerable Part)

You can have the perfect tube, razor-sharp focus, and G-code so good it should be on a TED Talk, but if your lens is scratched, or covered in mystery haze, your laser isn't cutting. It's just vaping on your material and hoping for the best. Lasers aren't cheap, and replacement parts even less so. Yes, even the CNC King has run a CO_2 beam through a fiber lens, I had to replace the entire head!

A clean lens gives you tight beams, crisp cuts, faster passes, and full power where it matters, on the material, not diffused like a sad flashlight. Less heat, less smoke, more wins with happy customers.

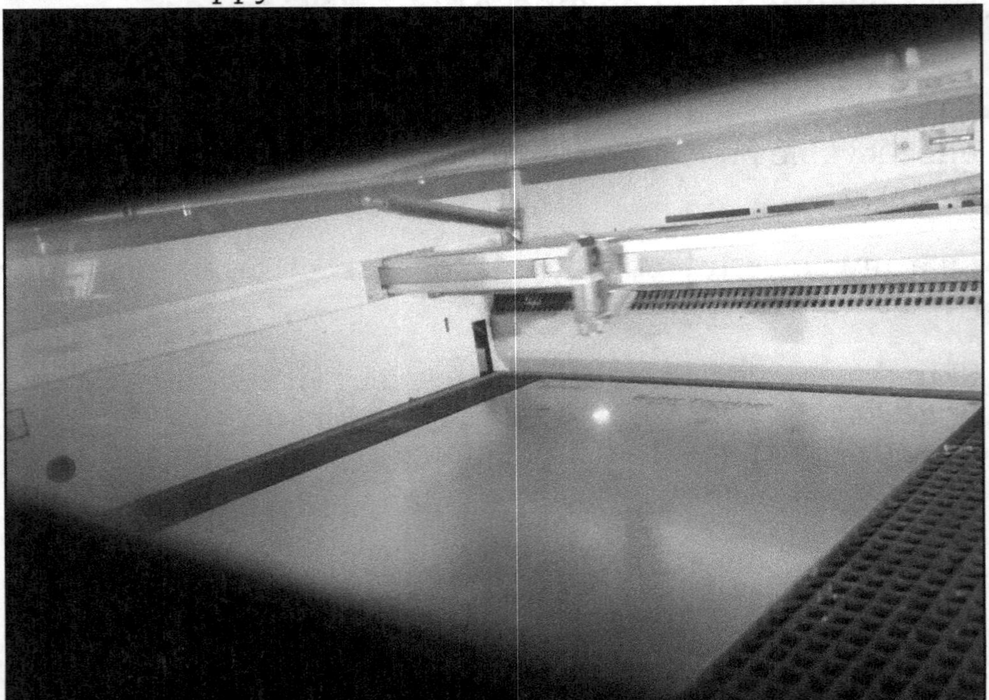

High utilization keeps this laser system consistently productive.

Dirty Lens Symptoms

Your machine might not scream, "my lens is gross," but it's absolutely trying to get your attention. Lasers are precision tools, and when that precision starts slipping, the lens is one of the first places to look.

Here are the subtle signs your optics are waving a white flag:

Dull Engraving

Text that used to be sharp and detailed now looks like it was scribbled with a half-melted crayon. Line work softens, edges blur, and suddenly your high-res vector file looks like it was engraved in low-effort mode. It's not the file, it's the beam losing focus through a haze of dust, resin, or even your fingerprints (when you handled the lens).

Charring Where There Shouldn't Be

Especially on light-colored woods, paper, or acrylic, your once-clean cuts now leave behind burn marks, edge staining, and unnecessary soot. And yet, you're cutting slower. That foggy lens isn't letting the full power through, so instead of slicing, your beam's just roasting the surface like it owes it money.

Excess Smoke at the Cut Site

A clean lens gives you a clean beam. Fast, focused vaporization. A dirty one? It diffuses the energy, turning your beam into a weak heat lamp that smolders instead of cuts. If you see smoke rising before the beam even hits the surface, you're not cutting, you're slow-cooking.

Inconsistent Depth

One pass cuts perfectly, the next barely leaves a mark. You start questioning your focus height, the material, the G-code. But in reality, it's the beam dancing through a blurry lens, sending energy wherever it pleases. I once had this issue, after investigation, it was a bug in the laser software.

Types of Lens Abuse

If you're seeing smudgy halos or soft, unfocused beams, congrats, you probably touched the lens with your bare hands. Clean it with a lens wipe. Brown or yellow film with patchy engraving? That's smoke stains from poor air assist and neglected cleaning. Hit it with an alcohol-based lens cleaner. Tiny craters in the glass? Those are pit marks from back reflections or debris flashback. That lens is done, regardless of cost, replace it.

Spidery fractures and beam dropouts point to cracks, likely from overtightening or heat stress. Time to replace it.

If the surface looks milky or foggy, that's cloudiness caused by too many dry wipes or bad alcohol. Get a new lens but don't drink the alcohol.

Cleaning Your Lens Without Breaking It

First, ditch the household hacks. No Windex. No vodka. And no spit! You need actual lens cleaner, formulated for optics though through the years, I've used Acetone in a spray bottle with great results as well at *CNCROi.com*.

Use lens tissue or proper optical swabs. A paper towel might feel soft, but under a microscope, it's basically a sandpaper delivery system. Before anything touches the lens, blow off loose dust with clean, dry, oil-free air (canned air works if held upright, no propellant spray).

This step matters more than you think as rubbing dust across the lens is how tiny scratches become permanent and over time, yet another reason to replace the lens.

When wiping, go gently and in a single direction. You're cleaning precision optics, not trying to win a floor-cleaning competition. A few light passes should do the trick. If it's still hazy or streaky, let it dry, then repeat. If it doesn't come clean after that, it may be damaged, not dirty.

Lenses are relatively cheap, but treat them like royalty and they'll give you a long life in return.

Maintenance Schedule

Regardless of your maintenance schedule, when the laser seems to lose power or the results don't match what I'm expecting, I assume the lens in gunky and check it out.

Material has a huge bearing on how often you clean the lens (and mirrors) as well as your air extraction system and air assist.

Pro Tips You'll Wish You Heard Sooner

Always install the lens the right way, concave side down (usually). When in doubt, ask your supplier, not the guy in the YouTube comments section. Use beam cones or nozzles to protect against splashback; it's cheap insurance that saves expensive optics.

After any crash, hard stop, or mysterious shop thud, check your lens alignment. Keep your backups labeled, "Good," "Emergency," and "Only If Desperate" are perfectly valid categories.

Clarity is Power

Your laser's only as powerful as the beam that actually makes it to the material. A perfect beam passing through a filthy lens is like yelling through a sock.

So here's the deal:
Keep the lens clean.
Don't overtighten it.
And stop touching the glass like it's a touchscreen.

And... and... this all holes 100% true for your mirrors as well, those tend to get dirty just as quickly so watch those too!

Next time your job runs like it's been demoted to "intern-level performance," don't jump straight to blaming the file, the settings, or the universe. Start with the one part that's been staring into the fire for hours: your lens.

It's not just a window, it's the whole view.

❑ Focus Your Beam, It's Q&A Time!

5.6 How do you select the right lens focal length for a job?

In the wild world of laser cutting and engraving, the lens you choose is kind of like picking the right glasses for your laser. Too short, and it's nearsighted; too long, and it's practically farsighted. Consult your manufacturer for guidance on the optimal lenses for your specific project, as there are too many variables to address comprehensively here. Use this information only as a rough guide.

Short focal lengths, typically around 1.5 to 2 inches (40 to 50 mm), are like your laser's reading glasses, super focused, and incredibly detailed. They concentrate the beam into a tiny spot, perfect for delicate work on thin materials like wood veneer and acrylic jewelry.

On the flip side, longer focal lengths of 4 to 5 inches (about 100–125 mm) act more like telescopes. They're less concerned with fine detail and more focused on reaching deep. These lenses spread the beam slightly, allowing it to maintain focus over greater distances. While you sacrifice some resolution, the advantage shows when carving or maintaining consistent etching (especially with fiber beams) on surfaces that aren't perfectly flat.

Using the wrong lens, however, will produce weak cuts, or weird tapers. That's why smart laser shops keep a collection of lenses on hand and swap them out based on the job, like a chef switching knives.

5.7 What role does air assist play in cutting performance?

Air assist is like your laser's hype man, blasting a focused jet of compressed air (or gas) right at the cutting point to keep things sharp and clean. Think of it as a tiny, invisible leaf blower with a mission: to blow away molten material, smoky fumes, and annoying debris before they can mess with the laser beam's path. Similar idea to a swirl ring on a CNC plasma but not as necessary or complex.

Without it, the laser's trying to do surgery in a steam room where heat builds up, edges get crispy, and if you're cutting wood, congratulations, you might've just started a tiny barbecue.

With air assist in play, the cut zone stays cool and drama-free. That means faster cutting speeds, cleaner edges, and none of that scorched, sad-birthday-candle look on your finished pieces. Air assist isn't new or innovative, my CNC plasma cutter uses it for optimal cutting, but it's especially critical for non-metals like wood, acrylic, and plastics, which are prone to melting or even catching fire. And in an enclosed environment, that's the last thing you want.

On metal-cutting lasers, plain air gets traded for the fancy stuff: oxygen for turbo-charged, combustion-driven cutting, or nitrogen for crisp, oxidation-free edges.

Not only does air assist make your cuts cleaner, it also protects your lens from catching stray goo and gunk, saving you from costly cleanups or replacements.

Expand globally while staying compliant with customs regulations.

5.8 How does pulse frequency affect engraving depth?

Pulse frequency, measured in Hertz (Hz), is basically how fast your laser goes "pew pew pew!" per second during pulsed operation. It's like the laser's personal rhythm.

Low pulse frequencies are like heavyweight punches: fewer hits, but each one packs serious energy, making them perfect for deep engraving or marking into tough stuff like metal, stone, or the occasional Terminator skull.

Crank the frequency up, and your laser trades brute strength for finesse, delivering a rapid-fire flurry of softer pulses that gently kiss the surface, ideal for delicate textures, fine details, or high-res images that whisper "art" instead of "blowtorch."

Metals tend to like it slow and powerful, while woods and plastics, depending on parameters, may enjoy the opposite.

Finding the sweet spot means balancing frequency, power, and speed like a laser DJ mixing a track. Too much of one, and your project can go from masterpiece to meltdown.

Nobody wants their engraving to look like it was done by a third rate student.

5.9 What materials require special coatings for laser marking?

Bare metals and certain plastics don't absorb laser wavelengths well, so trying to engrave them solo is like yelling at a brick wall. That's where laser marking sprays and pastes come in, they are the tattoo ink for surfaces that don't want one.

Apply the coating, hit it with the laser, and it bonds to the material, creating a sharp mark without deep engraving or damage. OK for stainless steel, anodized aluminum, and picky plastics or ceramics. Coatings boost contrast, reduce required power, limit heat damage, and help extend machine life. I prefer staying away from these at *CNCROi.com*, I just don't like the results or durability.

5.10 How do you minimize back reflections on thin metals?

Back reflections are the laser world's version of "I'm rubber, you're glue", except here, the comeback can fry your expensive optics.. When cutting shiny stuff like stainless steel or aluminum, the beam can bounce right back at the laser head, causing serious damage.

To avoid this laser boomerang, start with the right tool: fiber lasers handle metals way better than CO_2 ones. Next, dull the shine with blackening sprays or masking films, it's like putting sunglasses on your material. Tilting the workpiece a bit can also steer rogue beams away from the optics. And if you're fancy, some lasers come with reflection sensors that hit the brakes before disaster strikes.

5.11 What's the difference between raster and vector engraving strategies?

Raster and vector engraving are like the laser world's left and right brains, where one's an artsy sketchpad, the other's a precision ruler.

Raster engraving is the creative type: it moves back and forth, line by line, pulsing the laser on and off to "paint" your design pixel by pixel. This method shines for detailed images, shading, and grayscale work. Great for engraving portraits, logos, or textures on wood, acrylic, leather, or anodized metals. If you want your design to have nuance, depth, or look like a laser did an art degree, it's the winner.

Vector engraving (also called scoring) is the more technical sibling, replacing pixels with math. It follows the exact paths of your design like a laser-powered plotter, drawing straight lines, curves, or shapes in one smooth motion. It's ideal for outlines, crisp text, cut lines, and simple geometric patterns. Since it only hits defined paths instead of scanning an entire surface, it's much faster for clean, minimal designs.

Most projects benefit from both: raster for images and texture, vector for structure and sharp edges.

5.12 How do you manage beam alignment for multi-laser systems?

Multi-laser systems are like a band; if everyone's not in sync, the whole thing sounds (and cuts) like a disaster. Whether it's dual heads or combined beams, proper alignment is everything. Misaligned beams don't just look bad, they can cause weak spots, jagged edges, and can even fry your gear. In other words, not something to take lightly.

To keep the laser squad in line, you'll need alignment tools like mirrors, beam combiners, and targets. Use low-power test modes (a.k.a. "laser safe mode") to see where things land without melting anything. Regular cleaning of mirrors, lenses, and fiber connectors is also a must, as even a speck of dust can turn your beam into a firestorm.

For more complex systems, fancy alignment software or auto-calibration features take some of the manual pain away. I get my lasers serviced by a certified tech every year or two, it's a good insurance policy.

I love deep etching stainless steel with fiber lasers.

5.13 What's the impact of beam divergence on fine cuts?

Beam divergence is basically how much your laser beam can't keep it together over distance. Measured in milliradians (mrad), low divergence means the beam stays tight and focused; great for sharp details and tiny text. High divergence? That beam spreads like gossip, losing precision and turning crisp cuts into engraving lines.

For fine engraving, micro-machining, or anything that needs to look clean and pro, low divergence is your best friend. But if you're blasting out big, rough shapes or wide-area marks, a bit of spread isn't the end of the world.

To tame the divergence drama, you can tweak optics (like beam expanders), upgrade to higher-quality lenses, or use lasers with naturally low divergence, like single-mode fiber lasers that know how to stay in their lane.

5.14 How do you maintain optics to prevent power loss?

Laser optics are the eyes of your laser system, they are your lenses, mirrors, and windows, and if they're dirty or damaged, your machine basically needs glasses. Grimy optics scatter the beam, reduce power, crank up the heat, basically, the beam loses concentration.

To keep them happy, stick to a regular cleaning routine: use lint-free wipes, isopropyl alcohol or acetone (what I use but use whatever your laser manual swears by), and a gentle air blower. Never touch optics with your fingers as skin oil + laser heat = possibly a cracked lense (or mirror).

Check often for scratches, pits, or coatings that look like they've seen better days. Even the tiniest flaw can mess with beam quality. If it's damaged, don't be cheap, replace it. Running with bad optics can be overcome using more power, but eventually, you start melting things that are best kept cool.

Also, keep the beam path sealed and dust-free, and use proper air or gas purges to keep crud away. Don't be like me, having lasers between a plasma cutter and a routers.

5.15 How do you choose between galvanometer and gantry systems?

Galvo and gantry laser systems are like the sprinter and the bodybuilder of the laser world; both impressive, just with different strengths (and very different workout routines). One is built for speed; the other, for raw power.

Galvo systems use lightning-fast mirrors mounted on tiny, jittery motors to fling the laser beam across the material like it's doing precision parkour. We're talking hundreds of times faster than a gantry system. These are your go-to for small, high-detail jobs like barcodes, QR codes, serial numbers, crisp logos, or laser-etched memes on anodized aluminum keychains. If speed thrills and your parts are fun-sized, galvos are hard to beat.

They can knock out thousands of marks per hour with barely a sweat. But speed comes at a cost: galvos are limited in work area, usually under 12" (300 mm), about the size of a small pizza. Great for marking parts flying down a conveyor belt, not so great for cutting out things.

Gantry systems, on the other hand, are the steady, deliberate bodybuilders of the laser world. They move the entire laser head along X and Y rails like a robotic Etch A Sketch with a mission. Sure, they're slower, but they can take on huge jobs like thick materials, giant engravings, intricate cuts across meters of workspace.

You want to cut through MDF sheets, engrave large acrylic signs, or etch your logo into a coffee table the size of a mattress? Gantry's your ride-or-die. They're also better suited for tasks that require variable depth control or deeper cuts that galvos just aren't built to handle.

In short: Galvo = fast, flashy, and super precise for small stuff. Gantry = slow, strong, and built for big-time projects. *CNCROi.com* uses both, playing to each system's strengths.

It's like having a Formula 1 car and a monster truck in the same garage. Why choose, when you can have both?

5.16 What's the role of chiller systems in laser cutting?

Laser chillers are the unsung heroes of the laser world, basically the AC unit keeping your high-powered CO_2 or fiber laser from having a meltdown (literally). Lasers get hot fast, and without proper cooling, you're looking at thermal drift, fried components, and a very expensive paperweight.

Enter the chiller: it pumps temperature-controlled coolant (usually water or a water/glycol smoothie) through the laser, soaking up the heat and sending it off to chill-ville via compressors, pumps, and heat exchangers.

Good cooling means steady beam quality, longer component life, and fewer "uh-oh" moments during long jobs that were due yesterday.

Most chillers even have alarms or auto-shutoff features to stop things before they go full fire hazard. Small desktop lasers might get by with air cooling, but for anything industrial? A chiller may not be optional.

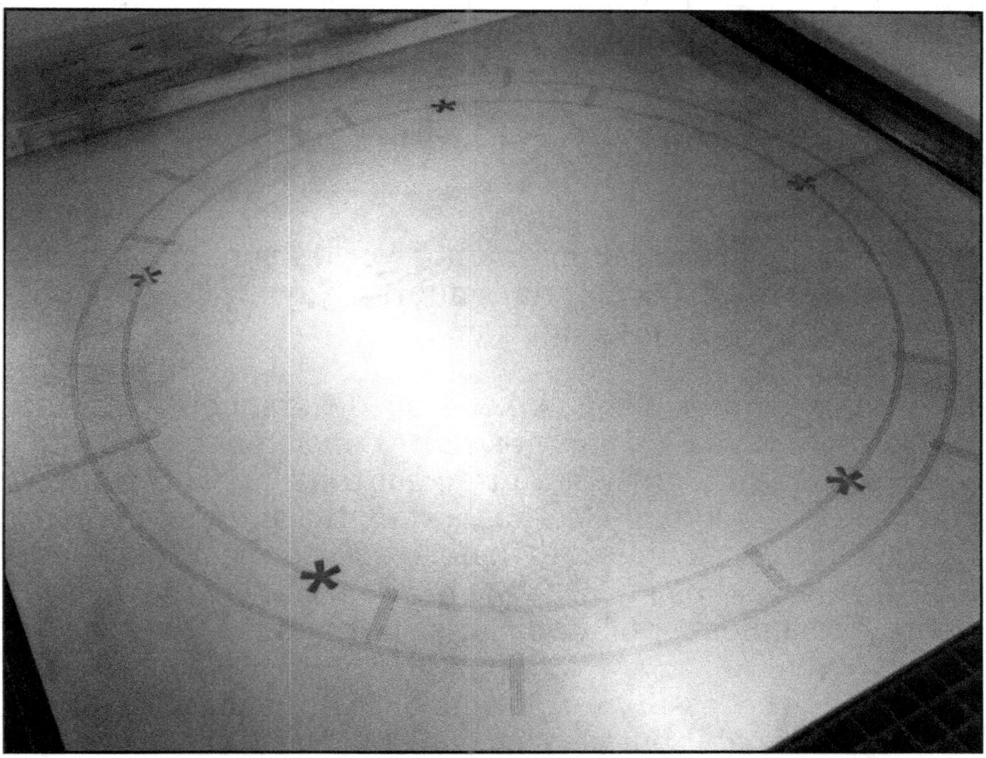

Use a gantry system for large work and a galvo system for small parts.

5.17 How do multi-pass strategies improve thick material cuts?

Multi-pass cutting is basically the laser equivalent of "don't force it, finesse it." Instead of blasting through thick material in one dramatic pass, you take several gentler swipes along the same path.

Why do it? First, it keeps heat buildup in check. One high-power pass can leave you with warped edges, or scorched surfaces. Multi-passing gives the material a breather between hits, leading to cleaner, more precise results.

Second, it lets lower-power lasers punch above their weight, which is perfect for shops that don't have a laser the size of a spaceship. It's also safer, with less risk of fiery surprises or rogue beam reflections.

The trade-off? It takes longer and your machine needs to be dialed in tight, if your alignment is off, those passes won't line up, and your "clean cut" turns into railroad lines. But when done right, multi-pass is a power move for cutting thick stuff that otherwise, you may not have the power to get through. Adjust beam height between passes!

5.18 How does surface preparation affect engraving contrast?

Surface prep is the laser engraving world's version of "don't skip skincare." If the material's not ready, your laser's just yelling at dirt, oil, and oxidation.

For metals, a quick clean makes a big differenceby ditching the grime and you get sharper, more consistent marks. Got a mirror-finish? Lightly scuff it to cut the glare and boost contrast. For wood and leather, a smooth sanding or a light finish helps engravings pop instead of disappearing into the texture. And plastics? Some need masking or surface treatments unless you like surprise burn marks.

This isn't as necessary for plasma but I do it anyways, partly out of habit but also because I like to work off as clean metal as possible, something you learn fast in welding.

5.19 What are the effects of laser polarization on cut quality?

Laser polarization is basically the direction your laser light prefers to "wiggle", and while that might sound like science trivia, it can actually impact your cut quality, especially in fine-detail or high-precision work.

For most CO_2 users, polarization doesn't matter much, as your laser does its job and everyone's happy. But with fiber lasers or certain optical setups, polarization can start acting like a picky chef: depending on the material and cut direction, it might subtly affect edge smoothness, marking depth, or create tiny grooves in places you didn't ask for.

In high-precision shops, especially those working with thin metals, microelectronics, or fancy plastics, some systems even let you rotate or control polarization to fine-tune results. For everyone else, just knowing it exists can help explain why that one cut looks weird when rotated 90°.

Polarization's not a big deal, until it is.

5.20 How do you safely manage fume extraction?

Laser cutting might look clean, but behind that pretty beam is a cloud of funky fumes, particulates, and mystery vapors, especially when you're zapping wood, leather, acrylic, or worse, PVC (which is basically chemical warfare). That's why fume extraction isn't optional, it's survival.

The gold standard is a proper fume extractor with HEPA and activated carbon filters. It sucks in the nasty stuff, filters out the toxic bits and smells, then either vents clean air back into the room or safely pipes it outside. Small lasers can use portable units; big industrial setups usually have ductwork that looks like a robot octopus, like you'd have for a CNC router's dust system.

Make sure your extraction system isn't underpowered, trying to vacuum a warehouse with a dustbuster won't cut it. And don't forget to change the filters regularly; a clogged filter turns your extractor into nothing more than a box.

Lasers are great for making custom cutting boards.

So beautiful you might not want to use them.

5.21 How does material reflectivity impact laser efficiency?

Some materials are just drama queens when it comes to lasers, where shiny metals like polished aluminum, copper, and brass love to reflect laser energy like they're holding up a mirror yelling, "Not today, photon!" That reflectivity makes them harder to cut or engrave, since most of the laser power just bounces off instead of getting the job done.

This causes two big problems: first, it tanks efficiency; you'll need more power, slower speeds, or multiple passes just to make a mark. Second, the reflected beam can bounce back into your optics or laser source and do some very expensive damage (a.k.a. "Oops, there goes $10K").

To fight back, pros often switch to fiber lasers, which have wavelengths that metals absorb better than CO_2. You can also cheat a little by applying dark coatings or laser marking sprays but as mentioned, I prefer to avoid that if at all possible.

5.22 What strategies improve edge finish on acrylic?

When treated right, it rewards you with gorgeous, glassy edges that look flame-polished straight off the bed. But mess up the settings, and it'll leave you with melty blobs.

First rule: use cast acrylic, not extruded. Cast melts clean and smooth; extruded cuts like it's in a bad mood, it tends to be rough, cloudy, and no fun to finish. Next, dial in your speed and power, as too slow or hot, and you'll char the edges; too fast or weak, and the cut won't make it through. A bit slower with slightly lower power often gives the smoothest results.

Keep your focus spot dead-on, ideally at the midpoint of the acrylic, I generally use a 2.5" (60 mm) lens for a tight beam. Use air assist sparingly; too much air cools the edge too fast and ruins that glossy finish you're chasing. Also, make sure your machine's motion system is smooth. Wobbles and vibrations turn pretty edges into squigglies.

5.23 How do you balance speed vs. power for efficient cuts?

Laser cutting is basically a game of Goldilocks: too slow and too powerful? You're roasting your material like it's on a campfire, expect charring, gooey edges, and kerfs so wide they look like you cut with a blowtorch.

Too fast and too weak? The laser politely grazes the surface, leaving behind half-cuts, ragged edges, and a strong urge to scream into a pillow.

The secret sauce? Finding that just-right balance where the laser delivers enough energy per inch to slice cleanly in a single pass. This magic combo depends on your material type, thickness, and laser power.

Thin acrylic? Think fast and moderate power. Thick wood or sheet metal? Slow it down and crank the juice.

It's a little like cooking, some things need a quick sear, others need a slow roast.

Most operators start with manufacturer-recommended settings, then run test cuts to dial things in, it's "guess and check," but with lasers and slightly more risk of combustion. The pros know it's worth the time, because once you find that sweet spot, everything runs smoother: cleaner cuts, faster jobs, and less stress on your machine.

Advanced laser systems even offer dynamic power adjustment, modulating the beam in real time as it moves through curves, corners, or inconsistent material. It's like cruise control for your beam, keeping everything steady and sharp.

Getting speed and power right doesn't just improve quality, it reduces material waste, cuts down on wear and tear, and saves on consumables.

Don't expect to dial things in perfectly on the first go. Even at *CNCROi.com*, there are fine adjustments made on the fly on a continuous basis.

5.24 How do you create grayscale effects in laser engraving?

Grayscale laser engraving is like giving your laser an art degree, where instead of just blasting lines, it learns to shade. In raster mode, the laser adjusts power and pulse frequency pixel by pixel: dark areas get more heat, light areas get just a gentle kiss.

To pull it off, start with a good image (300–600 DPI), then fine-tune your settings based on the material. On wood, more power = deeper burn = darker tones.

On anodized aluminum or coated metals, the laser tweaks reflectivity or coating thickness to create contrast without depth.

Engraving software like LightBurn or RDWorks makes life easier with built-in dithering and grayscale modes, just load the image, tweak the sliders, and let the laser do its moody masterpiece thing. Always run a test pattern first unless you enjoy surprises.

I perfer just doing black and white engravings as they provide clearer results in my opinion.

Turn retired saw blades into bold, industrial style clocks.

5.25 What emerging materials are expanding laser processing applications?

Laser tech isn't just about wood and acrylic anymore, it's getting into some seriously high-tech territory. We're talking advanced polymers like PEEK and polyimides, which are now being zapped into parts for aerospace, medical gear, and electronics thanks to their "survives-a-volcano" durability.

Carbon fiber composites and reinforced plastics used to be a nightmare to cut cleanly, but now lasers (especially ultrafast and tuned CO_2 types) handle them like pros. Perfect for making lightweight car parts and gear for athletes who hate gravity.

In flexible electronics, lasers are precisely patterning conductive inks and thin films, helping bring wearable tech and smart gadgets to life without tearing delicate materials. Science fiction, meet science fact.

Meanwhile, lasers are getting biological; shaping collagen scaffolds and bioresorbable polymers for implants and tissue engineering, because apparently lasers can now do surgery prep too.

And let's not forget ceramics. With the right lasers, even super-hard stuff like zirconia and alumina gets custom-cut for dental parts or mechanical wizardry. These materials are tough, but lasers are tougher, and smarter.

Yes, they are used for additive manufacturing as well.

5.26 How do you adjust focal offset for cutting vs. engraving?

Focal offset is where your laser decides to actually focus, and getting it right can mean the difference between a clean cut and a crispy mess.

For cutting, aim the focus point about halfway through the material. So, for 1/4" (6 mm) acrylic, set the focus around 1/8" (3 mm deep).

This keeps the beam tight through the whole thickness, giving you a clean kerf and minimal taper.

For engraving, keep the focus right on the surface to pack the most punch where it counts. Want crisp, sharp detail? Surface focus. Going deeper? You can tweak the focus just below the surface to help the laser chew through more efficiently.

Advanced users sometimes play with intentional defocus, slightly above or below; to soften edges, add shading effects, or blur harsh lines. I do this with lamacoids mostly.

5.27 What role does beam spot size play in marking resolution?

Beam spot size is basically your laser's paintbrush, and the finer the tip, the sharper the detail. A smaller spot means each laser pulse hits a smaller area, giving you crisper lines, finer shading, and high-resolution detail that would make even a tattoo artist jealous.

For example, a fiber laser with a spot size around 20–30 microns can pull off some wild precision, think microtext, barcodes, or tiny logos etched into metal parts. On the flip side, a CO_2 laser typically rocks a larger spot size (100–200 microns), which is perfect for general engraving, thicker lines, or slicing up wood and acrylic like a champ.

Spot size is influenced by a few key factors: the laser's wavelength (shorter is better for tight focus), the focal length of the lens, and the beam quality. A well-collimated beam that stays small and mighty across the entire work area means your fine detail looks sharp from edge to edge.

If you're working on ultra-fine applications like electronics, jewelry, medical tools, or those novelty grains of rice with names on them, investing in a system with a minimal spot size is non-negotiable.

Just keep in mind: the finer the detail, the more passes or slower speed you'll need. Generally speaking, high-res engraving is a marathon, not a sprint.

5.28 How can rotary attachments expand laser cutting capabilities?

Rotary attachments are how lasers learn to think in 3D. They let you engrave round stuff. Think tumblers, bottles, pipes, pens, whereby spinning the object as the laser does its thing, unwrapping your design like it's printing on a burrito.

Without a rotary? Expect wonky, stretched designs and a laser that loses focus faster than you on a Monday. With one? Perfect, crisp engraving all the way around.

There are two types: roller-style (great for smooth, light cylinders) and chuck-style (for heavy, oddly shaped stuff that needs a firm grip). Either way, it turns your flat-only laser into a 360° engraving beast.

Add a rotary, and now you're personalizing mugs, branding tools, or etching your logo on a steel pipe like a boss, all without buying a whole new machine. I never really got into that with *CNCROi.com* so the market is all yours!

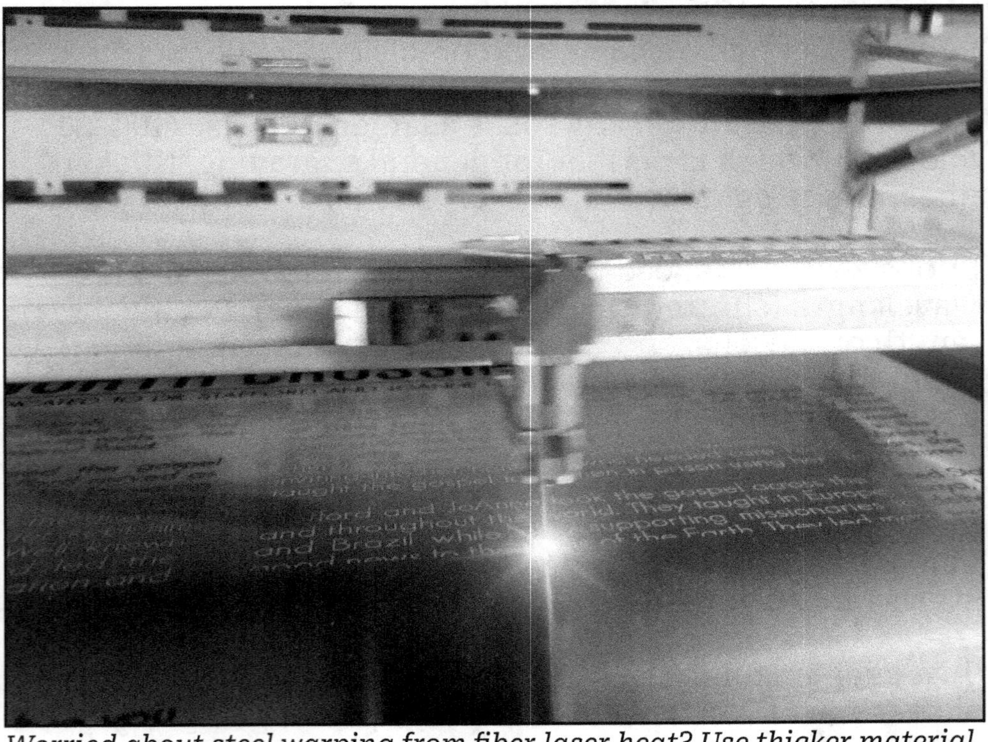

Worried about steel warping from fiber laser heat? Use thicker material.

5.29 How do you manage multi-material engravings?

A classic example of multi-material in a single "solid" product is using one laser beam for the metal and another for the wood, like a knife with a stainless steel blade and a wooden handle. Since these materials often sit at different heights, you need to account for that in your setup.

From my experience, it's usually more efficient to handle one material or height at a time. Trying to do everything in a single step can get confusing and mentally exhausting, especially as the number of units increases. While the idea of multitasking sounds great, a step-by-step approach is often better and helps ensure nothing gets overlooked.

5.30 What factors affect laser cut taper in thick materials?

Cut taper is that frustrating slant where your laser cut ends up wider at the top than the bottom, as if your material is rocking bell-bottoms. While perfectly vertical walls are the goal, thick materials almost always produce some taper due to how the laser beam interacts with the material. All CNCs produce a taper, just not to the same degree.

One key factor is the focus point: setting it at mid-thickness helps the beam stay effective throughout the entire cut depth, minimizing taper. If the focus is too close to the surface, the beam diverges as it cuts deeper, causing wider kerfs at the top.

The quality and focus of your beam also matter. A clean, well-focused beam concentrates energy precisely, reducing unwanted heat spread. Balancing cutting speed and power is crucial: moving too slowly overheats the top, melting and widening the kerf; going too fast means the laser doesn't fully penetrate, causing incomplete cuts and rough edges.

Choosing the right assist gas and pressure helps blow away molten material and prevent oxidation, keeping the kerf narrow and clean. For thick materials, a longer focal length lens gives a deeper depth of focus, helping maintain a straighter cut wall. Proper alignment and calibration ensure the beam path is straight and consistent.

5.31 How do you prevent scorching on light-colored woods?

Light-colored woods like maple, birch, and pine may look innocent, but they're scorch magnets when it comes to laser work, even a tiny burn sticks out like a stain on a white shirt. To keep your projects looking clean, start by masking the surface with painter's tape or transfer film.

This protective layer takes the heat so your wood doesn't have to, and you can just peel it off after cutting for a spotless finish. Next, optimize your speed and power. Go fast and use just enough power to get the job done as lingering too long means more heat, more burn. A strong air assist is also key; it blows away smoke and heat before it can settle on the surface. And don't forget to wipe the wood clean before starting as dust and resin love to amplify scorch marks.

With the right prep, clean settings, and a bit of airflow magic, you can engrave pale woods without leaving behind those telltale brown halos.

5.32 What's the importance of laser duty cycle on machine lifespan?

A laser's duty cycle is basically its work stamina; the percentage of time it can run at full power without overheating or throwing a tantrum. For example, a 70% duty cycle means the laser can go full blast for 7 out of every 10 minutes, then needs a 3-minute breather. Ignore that and you're pushing your laser into burnout territory.

Overdriving it can lead to thermal stress, worn-out components, and beam quality that slowly turns from good to bad. Industrial-grade lasers usually have beefy cooling systems and longer duty cycles, while hobby machines tire out faster. This is exactly the same concept as with welders where more expensive gives you more duty cycle time.

Know your laser's limits and plan smart where short runs, and breaks go a long way in keeping your beam strong and your repair bills low.

5.33 How do you balance pulse duration for deep engraving?

Pulse duration is how long your laser "blinks" each time it fires, and when it comes to deep engraving, that blink timing matters a lot. Short pulses hit hard and fast, blasting material away with minimal heat spread.

Great for metals, ceramics, and anything delicate you don't want scorched. Longer pulses, on the other hand, go deeper per shot but bring the heat, which can lead to melted edges or some unintended surface drama.

The trick is finding the sweet spot: for precise deep engraving, short, repeated pulses with moderate overlap help you carve clean layers without cooking your material. Softer stuff can sometimes handle longer pulses, but only if your power and speed are dialed in just right.

Test cuts are your friend for so many applications, they'll help you find that magic combo of depth, and detail.

5.34 How do external optics upgrades improve beam quality?

Upgrading your laser's external optics, lenses, mirrors, beam expanders, or collimators, is like giving your beam a gym membership and a personal trainer.

High-quality optics have better coatings, tighter tolerances, and smoother surfaces, meaning less distortion, scatter, and energy loss. Translation: sharper cuts, cleaner engravings, and more power where it counts. Beam expanders and collimators help keep your spot size small over longer distances. This is a must for big machines or ultra-detailed work.

Better mounts and alignment hardware also keep everything locked in place, so your beam doesn't wander mid-job. Sure, it's an investment, but one that pays off with faster cuts, prettier edges, and less material waste. If you're doing serious work, great optics give your laser a serious edge. You really do get what you pay for in the laser world.

5.35 What innovations are emerging in UV and ultrafast laser systems?

UV and ultrafast lasers are like the secret agents of the laser world, being ultra sleek, and precise.

UV lasers fire at super short wavelengths (~355 nm), making them perfect for micromachining delicate stuff like plastics, glass, and medical devices without frying the edges. Their "cold" ablation process is all finesse, no heat, which is ideal for electronics and high-precision parts where melting is not an option. Meanwhile, ultrafast lasers (like femtosecond and picosecond systems) deliver insanely short, high-energy pulses that vaporize material before heat even knows what's happening. That means you can machine sapphire, ceramics, or even diamond with surgical precision without burrs, burns, or cracks.

These tech marvels are showing up in everything from stents and semiconductors to quantum parts and micro-3D printing. Sure, they cost more than your average garage laser, but their precision is unmatched. It's on my buy list.

Acrylic hearts, keeping things organized.

5.36 CNC Laser: Top 10 Takeaways

Focal Length Selection: *Shorter lenses provide finer detail for thin materials; longer lenses cut deeper but reduce precision, ideal for thick or dense materials.*

Air Assist Function: *Clears debris, cools the cut, and protects optics, essential for clean, fast, and safe laser operation.*

Pulse Frequency Impact: *Low frequency enables deep, powerful engraving; high frequency creates smooth, shallow textures for fine detail.*

Coatings for Marking: *Some metals and plastics require laser marking sprays to achieve strong contrast without damaging the surface.*

Beam Alignment Needs: *Essential in multi-laser setups to ensure consistent cutting, prevent energy loss, and avoid equipment damage.*

Galvo vs. Gantry Systems: *Galvo offers speed for small areas; gantry provides size and power for large or thick materials.*

Chiller System Role: *Maintains stable laser temperatures, prevents overheating, and ensures consistent performance in long jobs.*

Multi-Pass Cutting: *Reduces heat buildup and improves cut quality on thick materials, especially when using lower-power lasers.*

Reflective Material Strategy: *Requires fiber lasers, surface treatments, or angled placement to reduce energy loss and protect optics.*

UV & Ultrafast Lasers: *Enable ultra-precise, cold processing for advanced materials in electronics, medical devices, and microfabrication.*

⊗ 5.37 Section Quiz

This quiz is like a CNC laser cutter, make the wrong move, and you'll burn through your material and ruin the whole job!

1. What does lens focal length affect in laser cutting?
   - A) Machine weight
   - B) Beam spot size and focus depth
   - C) Shield gas selection
   - D) Laser tube cooling

2. What is the purpose of air assist in laser cutting?
   - A) Cool the laser head
   - B) Reduce lens cleaning needs
   - C) Increase beam intensity
   - D) Blow away debris and improving the cut

3. How does pulse frequency affect engraving?
   - A) Controls depth and heat input
   - B) Determines engraving color
   - C) Changes laser wavelength
   - D) Reduces machine noise

4. Which materials need special coatings for laser marking?
   - A) Uncoated metals
   - B) Acrylic and MDF
   - C) Paper and cardboard
   - D) Plywood

5. What risk comes from back reflections for metals?
   - A) Increased power efficiency
   - B) Damage to laser optics or source
   - C) Improved edge quality
   - D) Reduced material cost

6. What's the difference between raster and vector engraving?
   - A) Raster is for outlines, vector is for filling
   - B) Raster uses higher power
   - C) Raster scans the surface; vector follows paths
   - D) There is no difference

7. How do you maintain beam alignment in multi-laser systems?
A) Realign optics regularly
B) Increase power settings
C) Use a shorter focal lens
D) Add air assist

8. What does beam divergence affect?
A) Engraving speed
B) Cooling water temperature
C) Fine-detail cut quality over distance
D) Dust collection efficiency

9. Why is optic maintenance critical?
A) Dirty optics reduce power and damage components
B) It changes the laser wavelength
C) It improves fume extraction
D) It raises table weight

10. What's the main advantage of galvanometer systems over gantry systems?
A) Lower price
B) Increased cutting power
C) Better for large-format cutting
D) Higher speed, especially for marking and engraving

11. Why are chiller systems used in laser setups?
A) Maintain optimal laser tube temperature
B) Cool the material being cut
C) Freeze the beam path
D) Reduce table vibration

12. Why use multi-pass strategies on thick materials?
A) Reduce tool wear
B) Lower power needs
C) Gradually cut through material without overloading
D) Change laser wavelength

13. How does surface preparation affect engraving contrast?
A) It doesn't matter
B) Clean, uniform surfaces improve consistent contrast
C) Rough surfaces always improve detail
D) Dirty surfaces reflect the laser better

14. What is the effect of laser polarization?
  A) Controls beam direction and affects cut edge
  B) Changes wavelength
  C) Increases beam intensity
  D) Eliminates the need for lenses

15. Why is proper fume extraction essential?
  A) To maintain beam alignment
  B) To protect operators and keep optics clean
  C) To reduce beam power
  D) To increase laser speed

16. How does material reflectivity impact cutting efficiency?
  A) Highly reflective materials absorb more energy
  B) There's no impact
  C) Reflective materials improve speed
  D) Reflective materials reduce laser absorption

17. How can you improve edge finish on acrylic?
  A) Increase air assist
  B) Use higher speed and moderate power
  C) Reduce focus distance
  D) Sand the surface before cutting

18. How do you balance speed vs. power for efficient cuts?
  A) Use maximum speed at minimum power
  B) Use maximum power at maximum speed
  C) Adjust power and speed to match material specs
  D) Always cut slowly for best results

19. What's required to create grayscale effects in laser engraving?
  A) Varying power and speed to adjust burn depth
  B) Only changing material
  C) Changing lens diameter
  D) Using air assist

20. What emerging materials are expanding laser applications?
  A) Only metals
  B) Only wood and paper
  C) Glass, ceramics, composites, and coated films
  D) Only plastics

Engraving 1/8" (3 mm) MDF ornaments.

Cutting usually goes faster than engraving.

✖ 5.38 Answer Key and Explanations

This answer key hits harder than a head crash on warped stock, but don't worry, the explanations cut cleaner than a well-focused laser on fresh acrylic.

1B – The focal length of a lens determines how tightly the laser beam can be focused. A shorter focal length gives a smaller spot size and shallower depth of focus, ideal for fine detail. A longer focal length provides a larger spot and deeper focus, better for thicker materials.

2D – Air assist helps remove vaporized material and smoke from the cutting area, preventing flare-ups and improving cut quality. It also helps reduce charring, especially in materials like wood or acrylic.

3A – Pulse frequency affects how often the laser fires per second. Higher frequencies can increase heat buildup, impacting engraving depth and texture. Lower frequencies give deeper, more distinct marks.

4A – Most metals (especially stainless steel and aluminum) require special coatings like CerMark or Thermark to create visible marks when using a CO_2 laser. Fiber lasers can mark metals directly.

5B – Back reflections from shiny or thin metals can reflect the laser beam back into the machine, potentially damaging the laser tube, mirrors, or focusing optics.

6C – Raster engraving moves line by line like a printer, great for images. Vector engraving follows paths defined by vector lines, used for cutting or outlining.

7A – In multi-laser or large-format systems, optical alignment is critical. Misalignment causes inconsistent power delivery. Regular checks and calibration keep beam paths accurate.

8C – Beam divergence describes how much the beam spreads out. A beam with low divergence maintains a tighter focus over longer distances, improving precision and detail in long or large cuts.

9A – Dust and debris on mirrors and lenses scatter or absorb laser energy, reducing cutting efficiency and potentially damaging optical components due to heat buildup.

10D – Galvanometer systems use fast-moving mirrors to direct the beam, making them much faster than gantry-based systems for engraving or marking. They're ideal for high-speed, small-area work.

11A – Laser tubes generate heat during operation. A chiller system keeps the tube within a safe temperature range, maintaining laser stability and lifespan.

12C – Using multiple passes helps cut thick materials without burning or stressing the machine. It allows controlled energy input, preventing heat damage or incomplete cuts.

13B – A clean and smooth surface allows even laser interaction, leading to more consistent engraving results. Dirt or texture can scatter or block the beam.

14A – Laser polarization can affect how energy is delivered to the material, especially in fine detail work or directional cutting. It's often considered in high-precision applications.

15B – Fume extraction removes harmful gases and particulates produced during cutting, protecting user health and preventing contamination of optics.

16D – Metals like aluminum and copper reflect much of the laser energy, making them harder to cut efficiently with CO_2 lasers. Special techniques or laser types (like fiber) are often required.

17B – Acrylic melts when cut. Using high speed and moderate power lets it melt cleanly without burning, producing a smooth, flame-polished edge.

18C – Optimal cutting involves balancing speed and power for each material and thickness to ensure clean cuts without burning or incomplete penetration.

19A – Grayscale engraving is achieved by modulating laser power and speed, controlling how deep and dark each area is burned into the material.

20C – New materials like ceramics, composites, and specialized coatings are opening up advanced applications in medical, aerospace, and electronics using laser processing.

Many woodworking tools work great on metal too.

Welding can achieve things that are impossible with wood.

Section 6
The Welding Booth
Dialing in AMPS to Avoid Cold Laps

Welding is, quite simply, the art of sticking metal together with controlled fire. Let's be honest, it's one of the coolest and most dramatic things you can do in a shop without accidentally blowing something up.

It's where sparks fly (literally), bright arcs dance, and molten metal fuses into shapes and structures that hold up bridges, buildings, cars, ships, pipelines, and even that rusty trailer you swear you'll fix someday.

At its core, welding joins two or more pieces of metal by heating them to the point of melting and allowing them to fuse together, sometimes with filler material, sometimes not, depending on the process.

There are a lot of different processes when it comes to welding, I've made my preferences clear in my previous book and why, but just to summarize, tack with FCAW + weld with SMAW is perfect for about 95% of the projects I do at *CNCROi.com*. Though since writing that book, I've grown to appreciate welding with with FCAW-S.

6.1 Why does industry love welding?

Because without it, half the world would fall apart. Ships, skyscrapers, aircraft, trains, oil rigs, heavy machinery, even your backyard BBQ grill all rely on welded joints to stay in one piece. Superglue isn't super all the time.

Welding creates strong, permanent joints between metal parts, often exceeding the strength of the base metal itself when done correctly. Ideally, the weld seam is stronger than the surrounding material, so if a failure occurs, it should happen outside the welded area, not within it.

Basically, if something big, metal, and important needs to exist, welding probably made it happen.

Welding isn't just about strength, it's creative, too. In any fabrication shop or metal art studio, welders shape and fuse metal into everything from sculptures to motorcycle frames.

Laying down a clean weld bead is deeply satisfying, the perfect balance of skill, patience, and amps. You're taming molten metal, after all. It takes time to learn, but once you get the hang of it, it's actually quite enjoyable and relaxing a process, and rewarding once the job is done!

Welding has its thrills: blinding arcs, sizzling sparks, hot glowing metal, and the sharp scent of burnt steel.

That's why welders suit up like armored knights. Safety first as "arc eye" (yes, a literal sunburn on your eyeball) is real, and trust me, you don't want it. I know plenty of welders who've been through it. For some, it's almost treated like a rite of passage, kind of like the "missing finger club" among woodworkers. But let's be honest: the membership dues far outweigh the benefits.

Turn scrap metal into stylish custom desktop organizers.

6.2 The Future of CNC Co-Bot Welding

CNC cobot welding merges the accuracy of CNC automation with the flexibility and safety of collaborative robots (cobots). As industries push for speed and consistency, cobots are set to redefine both large-scale manufacturing and small-shop fabrication.

Unlike traditional robotic welders, which are isolated and complex to program, cobots are designed to work safely alongside humans. With force-limited joints, intuitive interfaces, and built-in sensors, they're ideal for dynamic shop floors. When paired with CNC systems, these cobots can execute multi-axis welds with repeatable precision and minimal setup that even put industry veterans to shame.

In the near future, cobot welding systems will become smarter and more autonomous. Advances in AI and machine vision will allow them to recognize parts, adjust weld parameters on the fly, and learn from past jobs. They'll switch between GMAW, GTAW, or laser welding based on material and geometry, all within a connected digital workflow. This is already in practice in high-end shops where money isn't an ongoing concern.

For manufacturers, this means faster turnarounds, fewer defects, and scalable production without needing more skilled welders, which is a major advantage in today's labor-short market.

For smaller shops, it offers high-end capability without the complexity or cost of traditional automation. Cobots don't tire, don't rush, and don't call in sick, but they adapt and work through the night if needed.

Human welders won't disappear, they'll shift roles, becoming programmers, and creative problem-solvers. Cobots will handle repetitive or precision-intensive tasks, freeing people to focus on complex builds and finishes.

With better sensors and cloud connectivity, CNC cobot welding will bring real-time quality control and unmatched traceability. Leave the boring to the robots, not the welders.

6.3 Welding Safety Myths & Mistakes

It's gritty, it's satisfying, it's the backbone of modern industry, and it's also one of the easiest places to make dangerous safety mistakes if you're not paying attention.

So grab your helmet, zip up your jacket, and let's dive into 10 safety myths and mistakes about welding.

Myth 1: "It's just a little spark, no need for full PPE."

There are way too many YouTube videos showing someone tossing on a T-shirt, shorts, and sunglasses like they're ready to weld. Let's be clear, those "little sparks" are molten metal droplets flying at high speed, and they don't care that you just cleaned your sneakers.

Worse still are the high-intensity UV rays generated by the welding process itself. Sure, SMAW (SAW especially) provides some shielding, but even then, I saw fellow students in our College labs walking around with week-long sunburns in strange places coming from nothing more than small holes in their clothing or poorly positioned PPE.

Don't try to be a hero. Wear proper gear, and inspect it regularly. Like a router bit that's always just a little dull, welding clothes almost always have holes, it comes with the territory. Catch them before your skin pays the price.

Myth 2: "I can look at the arc for just a second; I'm fine."

Sure, and you can stare at the Sun for just a second too, right? Welding arcs put out blinding levels of UV radiation, and one careless glance can leave you with arc eye; a delightful condition where your eyeballs feel like they've been sandpapered from the inside out.

It doesn't hit you right away either, the pain shows up hours later, usually when you're trying to sleep. My late cousin Sylvain welded for 35 years and eventually went legally blind from tumors in his eyes. He had stories about how it happened, but they weren't the kind you'd want to replicate, let alone try to emulate.

Myth 3: "I don't need to ventilate, it's just metal fumes."

You mean the delicious cocktail of vaporized zinc, chromium, manganese, and nickel particles that can waltz straight into your lungs and throw a toxic party? Welding produces hazardous gases and metal fumes that build up fast, especially in enclosed spaces.

Ever heard of "metal fume fever"? It's like the flu, but with bonus lung damage. Crack a window, run an exhaust fan, or better yet, set up proper ventilation because coughing up metal-flavored goo isn't fun. Weld outside if you can!

Myth 4: "I'm just welding a little galvanized steel."

Welding galvanized steel without proper prep or protection is basically inviting a chemistry experiment inside your chest. When the zinc coating burns, it releases zinc oxide fumes, which, surprise, can give you metal fume fever real fast. Again, very preventable in most cases.

You might get chills, nausea, or feel like you partied too hard at a toxic metal concert. Always grind off galvanization in the weld zone and use a respirator or fume extraction if you're dealing with coated materials.

Myth 5: "The workpiece will ground itself."

Skipping the ground clamp because you assume the table is "probably grounded" is like assuming your parachute will probably open. You don't want to become the grounding!

Without a solid ground connection, you're asking for trouble: erratic arcs, weak welds, damaged equipment, or if you're really unlucky, a jolt of surprise 3-phase electricity.

Personally, I always ground the table by default, but most of the time, I take it a step further and clamp directly to the weldment, as close as possible to where I'm welding. It only takes a few extra seconds, and it does make a difference. It won't fix issues like using the wrong polarity, but it absolutely helps produce a more stable, controllable arc. This is doubly true if you are plasma cutting.

Myth 6: "I can weld near flammable materials if I'm careful."

This is the welding equivalent of saying, "I can juggle flaming chainsaws if I concentrate."

Sparks don't care how careful you are, they fly far (we're talking feet or meters), and they love sneaking into corners, under boxes, or into that pile of oily rags you meant to toss last week. And yes, they will wait until you leave the shop before lighting up your place like a disco party.

Clear your workspace of anything flammable, cover nearby combustibles, and keep a fire extinguisher within reach because you're a welder, not a firefighter.

Myth 7: "Gloves slow me down; I'll just weld bare-handed."

Enjoy the sensation of grabbing a torch handle that's been absorbing heat for ten minutes or casually brushing your hand across a still-glowing weld bead.

Sandblasting is a great way to pre finish metal projects.

Welding without gloves is a one-way ticket to the Burn Unit. Sure, gloves might cut your finger dexterity a bit (ok, a lot), but they also let you keep your fingerprints, which feels like a fair trade in my book.

That said, don't worry about missing out: even with gloves, you're still going to get burned and cut. Often. It's just part of the trade but at least it's more localized 3rd degree burns.

Myth 8: "A little UV exposure on my arms just gives me a tan."

Welding UV doesn't give you a tan, it gives you a radiation burn, minus the cool "green glow in the dark" bonus. It's the kind of burn that feels fine during the day, but by evening your skin is lobster-red, blistering, and so tender even a light breeze feels like sandpaper.

Welding sleeves, jackets, and neck covers exist for a reason.

Myth 9: "I don't need to check for gas leaks every time."

Leaky gas hoses or loose regulators can turn your workspace into a slow-motion disaster movie. Not only is wasted shielding gas expensive, but a sudden flare-up or pocket of flammable gas can turn your welding area into a fireball. This is one reason why I don't use gases at all at *CNCROi.com*, it's one less thing for me to worry about.

Check your fittings, hoses, and regulators regularly. Remember: explosions are cool in action movies, not in your custom fabrication shop.

Myth 10: "I've been welding for years; I don't need to follow every safety rule."

And here's the biggest myth of all: the overconfident veteran mistake. Experience is great, but complacency is the silent killer. Just because you've never been burned, shocked, or blinded yet doesn't mean you're immune.

Like a motorcycle driver, it isn't a question of if but when you get hurt, you want your PPE when it's your turn.

6.4 Lessons I've Learned about Welding

Welding is one of the most versatile and rewarding trades to be a part of, but the downside is how dangerous it can be. It's a high-risk craft, a constant dance with heat, light, electricity, and toxic fumes, and it will wear on you no matter how careful you are.

At Niagara College, watching my instructors weld day after day, one lesson became obvious: even the most experienced and safety-minded welders feel the toll over time. I saw the same thing firsthand later in a custom metal fabrication and welding shop.

You can't eliminate the hazards, but you can minimize them. And the welders who take safety seriously are the ones who come back ready to lay down another bead tomorrow.

No matter what anyone tells you, there's nothing truly healthy about welding. You're always trading a bit of your health for a paycheck, and it's important to keep that reality in mind. *I've written about my experiences as a jack-of-all-trades welder in Appendix G.*

Strong tacks help keep parts from warping while welding.

☐ Tighten your Arc, It's Q&A Time!

6.5 How does joint preparation affect weld strength?

Joint preparation in welding is like getting ready for a first date, if you show up covered in grime, misaligned, and totally unprepared, don't expect things to end with an invitation upstairs afterwards.

In welding, just like in romance, chemistry only works when you clean up first and put in some effort. A well-prepped joint is the equivalent of showing up with a clean shave and passionate conversation, as it sets the stage for a solid bond.

Skip the prep and you're basically forcing the weld to bond with rust, oil, paint, or whatever unidentified shop sludge you left behind. Not exactly a recipe for lifelong commitment. That's how you get porosity, inclusions, and cracking a week after you spent all that time welding. Clean surfaces mean the arc can actually do its job and penetrate properly, instead of dancing awkwardly over gunk like a welder at their first high school prom.

Edge beveling on thicker materials isn't just some fancy extra step for people with too much free time. Beveling helps the weld penetrate fully through the joint, giving you strength from the inside out. Without it, you're just laying down a nice-looking surface bead that might hold until someone sneezes on it too hard. Beveling also gives the weld pool a road to follow which makes your job easier.

And let's talk about fit-up. If your joint gap is too wide, your weld pool will fall straight through like it's on a slip-and-slide. Too tight? The fusion's incomplete, like trying to stir pancake batter with a toothpick. The goal is a precise fit, because a precise fit leads to a stronger, cleaner weld.

Joint design matters too. Whether it's a butt joint, lap joint, T-joint, or corner joint, each one distributes stress differently. Using the right design for the material and loading conditions make all the difference.

6.6 What is the role of preheating in welding thick sections?

Preheating is like giving metal a warm-up jog before the main event, especially if it's thick or high-strength. Without it, the weld cools faster than your coffee in a snowstorm, leading to stress hydrogen-induced cracking down the line.

Thick metal's a heat vampire, it sucks the warmth right out of the weld, leaving behind brittle, stressed-out microstructures that would snap under pressure. Preheating slows things down so the weld and heat-affected zone (HAZ) can cool evenly, boosting ductility and toughness. It's like professional bodybuilding, the slower your gains on-cycle, the slower you'll lose them off-cycle.

It also helps sweat out surface moisture, keeping hydrogen out of the mix, because no one wants tiny weld-ruining gremlins sneaking in. This becomes especially important for CWB (Canadian Welding Bureau) and AWS (American Welding Society) testing. For critical work, reheating isn't optional. You track it with temp sticks, infrared thermometers, or thermal cameras.

Say what you will about welding, but it has advanced dramatically over the past few decades, driven by modern science and increasingly rigorous specification standards.

6.7 How does post-weld heat treatment reduce residual stresses?

Post-weld heat treatment (PWHT) is basically a spa day for stressed-out metal. After all the chaos of welding, where it experienced rapid heating, cooling, and metal doing yoga against its will, things get tense. Residual stresses build up, leading to distortion and early retirement under load.

PWHT calms everything down. The metal is gently reheated (usually below transformation temps), held there to let internal stresses chill out, then cooled slowly.

PWHT reduces hardness, while boosting toughness, and restoring ductility so things don't break under pressure.

Temperatures and hold times depend on the metal's personality where carbon steel might need 600–650°C (1,112–1,202°F), while alloys can get picky and demand more precise treatment depending on their composition.

Skip PWHT on critical parts and you're basically leaving ticking time bombs of stress hiding in your welds. Not ideal for long-term success or your reputation as a welder.

6.8 What are the signs of undercutting in a weld, and how do you fix it?

Learning how to weld teaches you that everybody is a pro at undercutting, where you accidentally carve a trench into the base metal right along the weld toe. The result? A skinny groove clinging to your bead like an awkward shadow, quietly weakening the joint. This is so obvious that anybody can be taught to see and feel it.

You'll usually spot it as a sharp notch running alongside the weld, and it's not just an eyesore, it reduces the effective cross-sectional area of the weld and acts like a VIP pass for cracks, especially under fatigue or vibration under dymanic loading. Even if the weld looks solid, that groove can become a silent failure point just waiting for the right stress to strike, if your fingernail catches it, it needs to be fixed.

Undercutting usually happens when you're moving too fast, using too much arc voltage (amps), holding your torch at a weird angle, or weaving like you're painting a picket fence instead of laying metal. In other words, it's what happens when speed, heat, and sloppy technique form a terrible trio.

To fix it, slow down your travel speed so the filler has time to do its job. Drop the arc voltage a bit, or switch to a smaller electrode for more precise control. Adjust your torch or rod angle so the heat goes into the joint, not off into the void or onto the sidewalls. Remember, in most cases you want to weld at half of the included angle of the joint.

If the groove is already deep enough to matter, break out the grinder and lay down a clean, smooth repair weld.

6.9 How does interpass temperature control affect multi-pass welds?

Interpass temperature is the weld joint's "in-between" temperature. It's measured after one weld pass and before the next in a multi-pass operation. It's basically temperature tuning. Too cold and the weld turns brittle. Too hot and the structure softens up. Nail the sweet spot and everything comes together.

Controlling interpass temp affects everything from microstructure development to stress control and overall weld quality. If it cools too fast between passes, you might end up with brittle phases or invite hydrogen-induced cracking. But if it gets too hot, grain coarsening and toughness loss crash the party.

Welders keep tabs on interpass temp using contact thermometers, infrared sensors, or pyrometers , not just by touching the metal and going, "Yep, toasty." The acceptable range is usually laid out in the welding procedure specification (WPS), and some alloys have strict upper limits to maintain strength, corrosion resistance, or both.

Nailing interpass temperature control means each weld layer bonds properly, the structure develops consistently, and you avoid sneaky defects that could turn into big problems down the road.

Keep in mind, for the vast majority of projects, this isn't much of an issue because most welding is done in one pass and the results aren't critical to the application.

6.10 What are the differences between AC and DC welding currents?

AC (alternating current) and DC (direct current) are like the Coke and Pepsi of welding power sources where both do the job, but each has its quirks and die-hard fans.

DC welding keeps things simple: the current flows in one direction, giving you a steady, smooth arc that's easy to control.

It's like cruise control for your welds. You get easier starts, less spatter, and deeper penetration. What more could a welder want? DC is the go-to for SMAW, GTAW, GMAW, and MCAW welding on materials like steel, stainless, and copper alloys. Plus, you get options: DCEN (direct current electrode negative) if you want to melt into the base metal like butter, or DCEP (electrode positive) if you need some surface cleaning action.

AC welding, on the other hand, can't make up its mind, it switches direction 60 times a second (in North America, anyway). It's like being married. But that's exactly what makes it useful. If you're working with magnetic materials like cast iron or certain steels, DC can throw a tantrum and cause arc blow where the arc gets deflected by magnetic fields. AC steps in like a therapist, calming things down by constantly reversing polarity.

It also really shines with aluminum GTAW welding. Aluminum has a stubborn oxide layer that's harder than the metal underneath with a higher melting point. DC would rather pretend that oxide doesn't exist, but AC punches through it with a built-in cleaning effect, giving you a clean, shiny weld.

I enjoy SMAW using 7018AC but DC is easier for a beginner.

Every scrap has a future project waiting.

6.11 How do you select the right filler metal for dissimilar metal welding?

Welding dissimilar metals is like matchmaking picky celebrities, each has their own quirks: different melting points and thermal tantrums. To keep the relationship from cracking (literally), pick a filler metal that plays nice with both, usually siding with the weaker one for peace.

Think of it like: carbon steel meets stainless steel, and ER309 shows up as the chill mediator with corrosion resistance and expansion compromise.

Actually read the instructions, follow the WPS gospel, and always consider the joint's future. Will it face corrosion, or cold temps? Do test welds first. Better to find out your couple can't stand each other before the wedding when they are bonded for life.

Don't be shy to get advice from the filler supplier, they live and breath this knowledge, use it to your advantage.

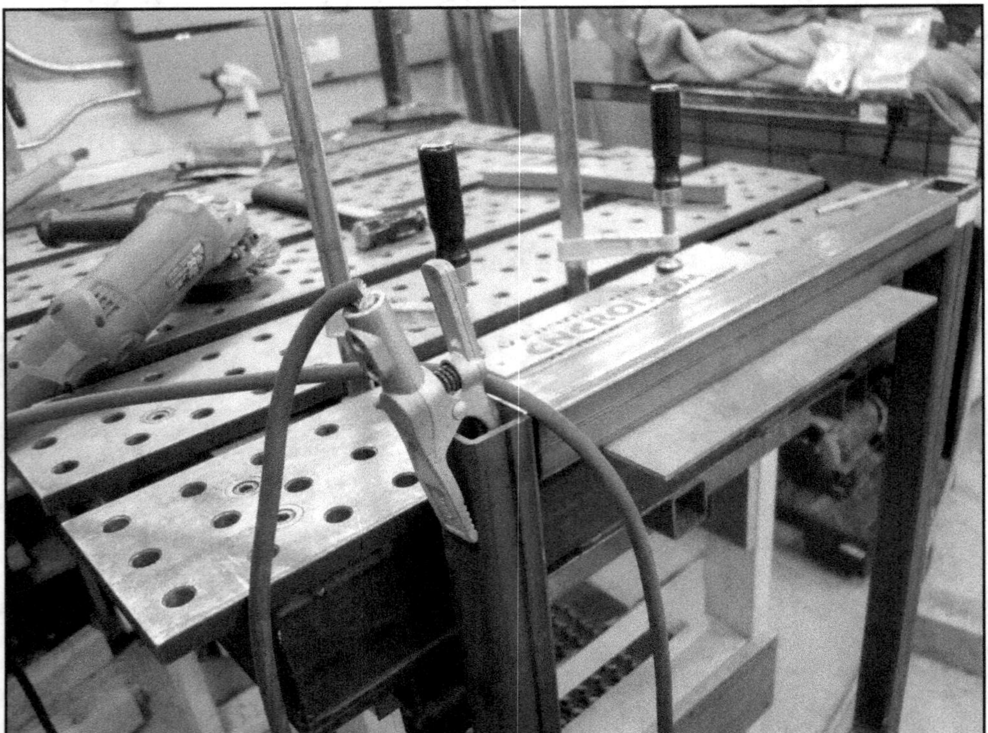

Clamp your ground to the part to be welded whenever you can.

6.12 What is porosity in welding, and what causes it?

Porosity is the acne of welds. Tiny gas bubbles get trapped in the metal, leaving it weak, and about as trustworthy as a wet paper bridge. Not great for anything important, like structural supports or space travel.

The usual suspects? Dirty metal, bad shielding gas coverage (thanks, wind), sloppy settings, or E7018 electrodes that were accidentally dipped in water because the welder thought they were E6010s. Basically, if your prep is lazy or your gear's a mess, porosity will become your best friend.

Want to avoid it? Clean everything, bake your rods (if E7018), dial in your machine, and keep that shielding gas flowing like it's blowing birthday balloons.

Got porosity? Grind it out and re-weld. Every weld deserves a second, or third chance. Bubbles have no place in a weld.

6.13 How can you identify and prevent crater cracking?

Crater cracking is the weld's awkward goodbye. Stop too fast, and the molten metal freezes into a little pit that shrinks and cracks under pressure, like a stressed-out student during ticket testing.

It usually shows up as a tiny radial crack right at the weld's end, waiting to cause trouble later. Not cute, and definitely not harmless.

To avoid crater cracking, always backfill the crater before breaking the arc, or use your welder's crater-fill function.

Another solid move is to weld just past the end point and gradually taper off, like a graceful exit instead of a mic drop. And seriously, no ghosting. Pulling the electrode away too quickly starves the weld, it isn't receiving enough molten filler metal or heat to properly fill and fuse the joint.

If a crater crack does rear its ugly head, grind it out completely and re-weld the area. Even the tiniest crack can become a point of major failure under dynamic stress.

6.14 What are the main advantages of pulse GMAW welding?

Pulse GMAW (or pulsed spray transfer) welding is like giving your welding machine a heartbeat, it rhythmically alternates between high and low current levels, creating a smooth, controlled droplet transfer from the electrode to the workpiece. This process brings some pretty slick advantages over your grandpa's old-school spray or short-circuit GMAW methods:

Reduced heat input: Because the current pulses instead of blasting nonstop, the average heat is lower. That means less metal warping, less heat-affected zone (HAZ) drama, and fewer surprises when you're welding thin stuff that usually acts like it's made of tinfoil.

Spatter control: Those pulsed droplets behave themselves, no wild arcs, no fireworks show, just smooth, stable metal transfer. Less spatter means less grinding and cleanup, and more time pretending you're working when you're really just watching the puddle sparkle.

Out-of-position welding made possible: With lower heat and a more focused arc, you can weld vertically, overhead, or even while hanging upside-down if you are so adventurous. Regular spray transfer would just drip molten chaos while pulse GMAW gives you control.

Enhanced penetration and fusion: The high-current pulse phase dives deep into the joint like it's doing metal surgery, while the cool-down phase prevents it from melting everything around it. It's like getting a tattoo with an ice pack, being both painfully precise, yet surprisingly gentle.

Sure, these pulse GMAW machines are more complicated than a remote control and definitely cost more than a six-pack, but they bring serious bang for your buck in industries like automotive, aerospace, and high-end fabrication, where clean welds and repeatable quality are more important than saving a few bucks on gear.

Remember, SMAW can do a lot of what pulse systems can for smaller shop, just a little slower and more relaxing.

6.15 How does shielding gas composition affect weld penetration?

Shielding gas in welding is like sunscreen for your weld pool, it blocks unwanted elements like oxygen and nitrogen from crashing the party and messing up your welds. But it's not just protective; it also changes how hot, deep, clean, and good-looking your welds turn out.

If you're doing GMAW welding, pure argon gives you a smooth arc and shallow welds, which is perfect for aluminum, but not ideal for steel. Mix in a little CO_2 or O_2, and you get better penetration and smoother flow on steel. Go all-in with 100% CO_2 for budget-friendly, deep welds, just brace yourself for a spatter explosion.

For GTAW welding, pure argon is your everyday workhorse. Need more heat and faster welds on thick materials? Add helium to the mix. And if you're welding stainless steel, a touch of hydrogen or nitrogen adds extra shine and penetration power.

6.16 What safety precautions are critical when welding galvanized steel?

Welding galvanized steel? Congrats! You're about to turn a solid zinc coating into toxic fumes! Yes, I've welded galvanized steel. It's miserable to weld and even worse to clean up. Definitely not something you volunteer for. Thanks, Jeremy, for letting me give it a try. Deep coughs.

Galvanized steel is coated with zinc to fight rust, but weld it without caution and you'll inhale zinc oxide. Welcome to metal fume fever, a delightful combo of chills, fever, nausea, and feeling like you got body-slammed by the flu.

Ensure you have proper ventilation, it should already be a given when welding. Always wear full PPE, including a respirator. Grind off any zinc coating before welding to reduce smoke and achieve a cleaner, stronger weld.

Avoid working in confined spaces unless you've got serious safety protocols in place, or better yet, just don't do it.

6.17 How do you calculate the weld size needed for a structural application?

When it comes to structural welds, you're basically matching muscle to mayhem, making sure the weld holds strong when things get real and doesn't tap out before the base metal does.

Here's the fast and filthy version: First, figure out what kind of load you're dealing with; shear, tension, bending, or the full stress buffet. Then, check your materials, because both the base metal and weld metal strengths matter.

For fillet welds, calculate the throat thickness using 0.707 × leg length (yep, Pythagoras makes a guest appearance). Multiply that by the weld length to get the cross-sectional area, and make sure it can take the applied force without blowing apart like a cheap lawn chair.

Weld layout also matters, more welds aren't better if they're placed like a toddler playing with glue. Follow your code rules if you care about liability, and don't skip safety factors unless you're aiming for a very short career.

Too small, and the weld could fail spectacularly. Too big, and you're just burning time and electrodes.

6.18 What are the visual inspection criteria for weld acceptance?

VT (Visual Testing) is often the first and sometimes only line of defense against weld disasters. Think of it as a weld's job interview: if it looks bad, it probably is bad. You're looking for cracks, craters, or porosity trying to hide in plain sight. The bead should be smooth and even.

There shouldn't be any wild undercut, overlap, or spatter unless chaos was somehow part of the design. The weld's size and length need to match the spec. Close enough will just generate a pink slip. And finally, it should tie cleanly into the base metal, with no SMAW arc strikes, missed starts, or half-hearted endings.

Tools help like gauges, fillet rulers, maybe a magnifier if you want to look real serious. Codes like AWS D1.1, ASME Section IX, ISO standards, or CWB requirements tell you exactly how picky you need to be and in welding, unlike the good old days, the details matter for liability sake.

Even with advanced NDT (Non-Destructive Testing) methods like UT (Ultrasonic Testing), RT (Radiographic Testing), MT (Magnetic Particle Testing), or PT (Penetrant Testing), VT (Visual Testing) still catches surface issues early, and early fixes mean fewer headaches (and less paperwork). Always VT inspect your welds first.

6.19 How do automated and robotic welding systems improve productivity?

Automated and robotic welding systems have significantly transformed industrial welding by delivering speed, consistency, and repeatability that manual welding struggles to match. Unlike human welders, robots can operate continuously without fatigue, maintaining precise travel speeds, torch angles, and bead placement across long production runs or multiple shifts.

This consistency reduces weld defects by eliminating human variability, leading to fewer rejected parts, less rework, and minimal scrap. Robotic systems offer tight control over key parameters such as arc voltage, wire feed speed, and travel speed, ensuring each weld meets strict quality standards.

In high-volume industries such as automotive, aerospace, heavy equipment, and structural fabrication, robotic welding cuts labor costs and cycle times, boosting productivity and profitability. These systems also enhance workplace safety by handling hazardous tasks, including welding in confined spaces and reducing exposure to heat, fumes, and arc radiation.

Robotic welding demands upfront cost and skilled support, but when applied to the right tasks, it increases output, improves quality, and lets human welders handle more complex work.

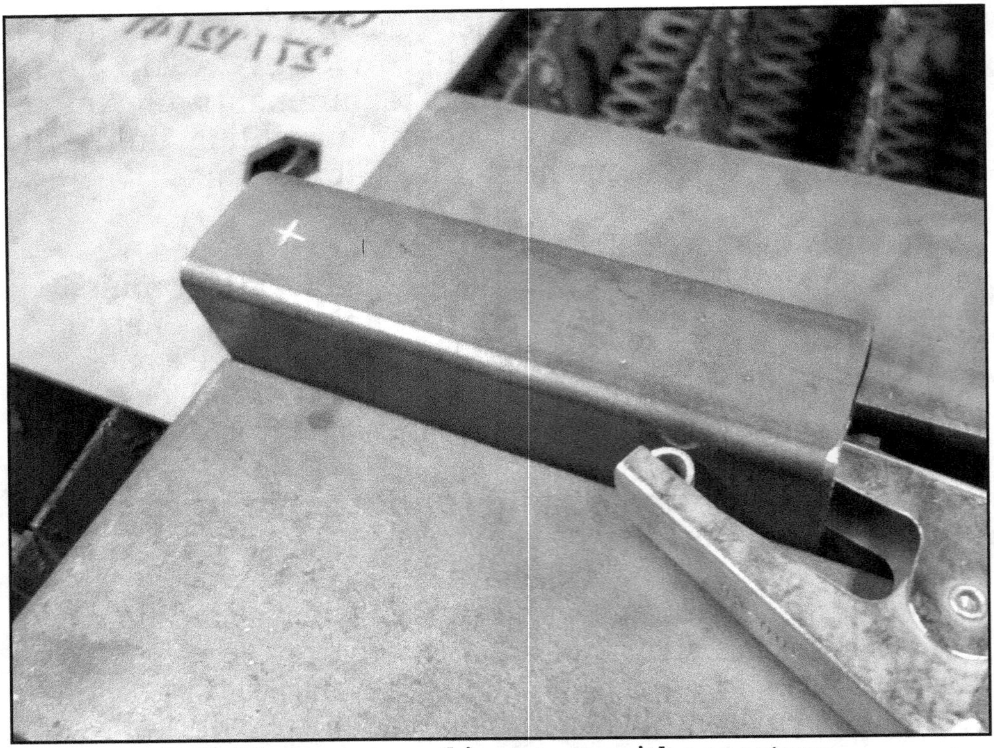

Reliable layout marking starts with soapstone.

Cold cut chop saws provide faster, cleaner cuts than abrasive saws.

6.20 What are the key parameters in laser welding vs. traditional arc welding?

Laser Beam Welding (LBW) and traditional arc welding (GMAW, MCAW, GTAW, SMAW) both join metal, but the way they do it is like comparing a lightsaber duel to a barroom brawl.

Laser Welding: Precision Death Ray Mode

Imagine a laser-guided scalpel from the future, but instead of healing, it vaporizes metal with pinpoint vengeance. Laser power is your zap level: crank it up to slice steel like foam, tone it down to whisper sweet nothings into aluminum. Just focus the beam right, or you'll host a tiny, expensive rave on your workpiece.

Speed matters, too fast and your weld vanishes, too slow and you crater it. Shielding gas like argon keeps things pristine, because even welds have standards. Want flair? Go continuous (power ballad) or pulsed (techno rave). The payoff? Razor-thin, deep welds with the distortion of a meditation playlist.

Arc Welding: Old-School Brawler

Enter arc welding, the seasoned street fighter of metalwork. No frills, just fire. Crank the amps for more oomph, it's the "punch harder" button. Voltage sets your arc's reach and your bead's swagger. Travel too fast? All sparks, no depth. Too slow? Welcome to molten fondue. Electrode choice is like picking your weapon in a video game; mess it up and pay the price. Shielding gas or flux? That's your invisibility cloak against contamination.

Arc welding doesn't need lasers or sci-fi flair. It works in the rain, eats rust for breakfast, and fixes farm gear without blinking.

Laser welding is the stealth assassin: sleek and surgically precise. Arc welding is the burly brawler: loud and gritty. Either way, metal melts, depends if you want startup vibes or a steel mill soundtrack.

6.21 How do you prevent hydrogen-induced cracking in high-strength steels?

Hydrogen-Induced Cracking (HIC), aka cold cracking, is like the sneaky ninja of welding defects, it doesn't strike right away, but when it does, it shows up uninvited and wrecks everything down the line. It happens when hydrogen atoms worm their way into your weld, lurk in high-stress zones, and then form cracks after the weld cools. This process may even only appear weeks later!

You think the job's done, and then your shiny weld says, "Surprise, I'm broken now!" To keep this hydrogen drama from ruining your life, you've gotta play defense. Use low-hydrogen electrodes or filler metals, basically the equivalent of putting your weld on a strict no-junk-diet. Preheat the base metal like it's a cold burrito you want evenly warm inside, as this slows down cooling and gives hydrogen time to escape before it settles in like a squatter.

Keep interpass temperatures in check, because stopping mid-weld and letting it chill is basically inviting hydrogen back to the party. A post-weld heat treatment (PWHT) is like a spa day for your weld, it relieves stress and lets trapped hydrogen diffuse out peacefully. Treat your E7018 consumables like royalty: keep 'em dry, don't let them touch damp, rusty grime, and absolutely clean your base metal, this cannot be emphasized enough.

6.22 What are the effects of welding on the microstructure of metals?

Welding isn't just about melting metal and hoping for the best, it's basically a high-stakes makeover for your material's microstructure. When you blast a section of metal with intense heat and then cool it down like it just ran a marathon, the internal structure goes through some serious drama.

In carbon steels, if you cool too fast, the heat-affected zone (HAZ) throws a tantrum and turns into martensite, which is a super-hard but super-brittle phase that's basically metal's version of snapping under pressure.

Alloy steels, meanwhile, might lose their nice, tempered vibe and go full weak-sauce on toughness.

Stainless steels get extra sensitive and might start crying chromium carbides at their grain boundaries, causing a sad condition called sensitization, which is a fancy term for "rust magnet."

Even the weld metal itself isn't immune; as it solidifies, it often forms chunky, coarse grains that aren't as tough as the base material. Whether your weld turns out heroic or horrible depends on things like cooling rate, alloy content, and how you babysit the whole process. That's why smart welders don't just melt and hope; they preheat, control interpass temps, slow down cooling, and sometimes give the weld a post-heat spa treatment (PWHT).

By the way, when I mention "grain" in metal, I'm referring to a microscopic structure that can't be seen with the naked eye. It relates to FCC (face-centered cubic) and BCC (body-centered cubic) crystal arrangements at the atomic level, which are invisible without specialized tools. Because of this, grain structure is not something you can evaluate through visual inspection (VT) after you are done welding unless you are Superman.

A manual plasma torch punches holes fast, but they can be rough.

6.23 How do you perform a bend test on a welded joint?

A bend test is like putting your weld through a yoga class where it's forced into a full 180° backbend from both the face and the root. The goal? No cracking under pressure!

First, you slice a sample strip from the weld. Then, you place it in a bend tester, basically a metal torture device, and apply force in the middle until it folds like a taco. Once it's bent, you check for cracks, gaps, or any embarrassing flaws that suggest the weld wasn't ready for prime time.

If the weld survives without splitting its seams, congrats; it's flexible, and free of major defects. It's a cheap but brutally honest way to find out if your welds can handle the real world. This is also fantastic way to learn how to perfect your welding abilities, I learned so much doing this.

6.24 What is the purpose of a backing bar in welding?

A backing bar is like a trusty safety net for your weld, catching the molten metal before it belly-flops through the joint and falls to the floor or worse, inside your welding boots. Don't ask me how I know this.

When you're welding from one side and can't access the back, that molten metal needs something solid to rest against. That's where the backing bar steps in, it's a dependable sidekick made of steel ($), copper ($$), or ceramic ($$$) that prevents your weld from dripping out.

Ceramic and copper bars are the VIPs of the backing bar world. They don't stick to the weld metal, can handle scorching temperatures, and help keep things neat and smooth at the root. Steel backing bars are tougher and cheaper, but they might need to be removed afterward to avoid defects or corrosion issues.

Sometimes the bar becomes a permanent part of the joint, while other times, it's just there for backup duty and gets booted once the weld cools. Either way, using a backing bar helps nail consistent root passes, reduces burn-through, and gives your welds the solid foundation they deserve.

6.25 How does submerged arc welding (SAW) differ from gas-shielded processes?

SAW is the introvert of the welding world, it works hard under a blanket of granular flux, skipping the sparks and flash for quiet, heavy-duty results.

What makes SAW special? It delivers metal like a firehose compared to the garden hose pace of other processes, laying down thick welds fast. The arc stays hidden under the flux, so there's no blinding light. It thrives in automated setups, making it perfect for long, repetitive welds in mass production.

It pairs best with carbon and low-alloy steels but isn't into aluminum or anything too thin. And don't expect it to experiment with positions, it's strictly flat and horizontal, as of the writing of this book.

Gas-shielded processes such as GMAW and GTAW move with the agility of gymnasts. SAW is the powerlifter of the welding world, understated yet engineered to take on massive fabrication whenever the joint is flat and gravity becomes an ally.

6.26 What is the role of a welding procedure specification (WPS) and how is it developed?

A WPS is the official playbook for welders, a precise set of instructions saying, "Here's how not to mess this up." It covers everything: what metals you're welding, which filler to use, how to prep the joint, what temps to hit, and exactly how to run the weld. No guesswork, no creative flair, just the welding gospel, written in amps and millimeters.

Why so serious? Because without a WPS, welders might improvise like jazz musicians with blowtorches, and that's how bridges crack and tanks explode. A WPS keeps everyone on the same page and liability in check.

But you don't just make one up. First comes the test weld, a Procedure Qualification Record (PQR), where the proposed process is trialed and then poked.

6.27 How do you handle welding thin materials without burn-through?

Welding thin metal is like toasting marshmallows with a flamethrower, one wrong move and you've got a gooey mess (or in this case, a hole). The trick is all about heat control: lower the amps, and speed up your hand.

Pulsed GMAW or GTAW is your best friend here, delivering heat in polite little bursts instead of one big inferno. Small wires, fast travel, and sometimes a copper backer (a heat-sucking sidekick) help keep things cool. Stitch welding or a bunch of tack welds also spread the heat love around instead of cooking one spot to death.

At the end of the day, even with the right settings, it's all in the wrist. Smooth hands, steady travel, and practice are what keep your weld from turning into Swiss cheese. Make sure to have your body solidly anchored as well, it makes a huge difference in all positions, thick or thin metal welding.

6.28 What is the importance of controlling weld bead overlap?

Weld bead overlap happens when molten metal lazily oozes over the base material without actually fusing to it, like a pancake sliding off the griddle before it's cooked. It might look like a weld, but underneath it's just a weak, unfused flap that screams, "I'm gonna crack later!"

This sneaky defect doesn't just hurt the weld's strength, it also sets the stage for stress buildup, corrosion, and early failure. Visual inspectors spot overlap quickly, and they're not impressed. Expect your weld to get the red pen treatment and a trip back to the grinder.

Preventing overlap means dialing in the heat (amps), hot enough to fuse, not so hot it floods. Keep your travel speed steady, angle your torch like a pro, and don't lay down more filler than the joint can handle. With the right technique, you get a smooth, solid weld that's both structurally sound and inspector-approved. Less rework, more high-fives.

6.29 How can distortion be controlled during welding large assemblies?

Welding distortion is metal's way of saying, "You heat me unevenly, I bend how I want." It's caused by parts expanding when hot, then shrinking as they cool, like a cross-wind happening during a FMD 3D print. One moment your perfectly aligned assembly looks great, the next it's done and you can just watch the failures start to mount.

Left unchecked, distortion turns big projects into big problems. In large structures, a little warping here and there stacks up like bad decisions in a group project. Suddenly, your frame doesn't fit, your flanges don't align, and someone's yelling about tolerance violations.

To keep metal in line, you have to outsmart it. Weld in a balanced sequence; if you heat one side, hit the other before it gets jealous. Clamp your parts down like they owe you money. Use less heat, weld faster, and if you're really thinking ahead, pre-bend things so they warp into the right place. Skip welding, backstepping, and controlled cooling are all tricks in the distortion wizard's toolkit.

Not sure about the fit? Use a soft tack you can still move.

6.30 What are the environmental impacts of welding fume emissions?

Welding fumes are like the sketchy side effect of an otherwise cool process, a smoky cocktail of metal oxides, silicates, and invisible cancers that your lungs don't need.

In the shop, these fumes can cause everything from coughing fits to chronic illness. Out in the wild, fumes release pollutants like ozone and nitrogen oxides, and toxic metals like chromium or manganese can crash the local ecosystem party. I'm sure the UV exposure also releases other fun chemicals within the local environment as well.

The fix? Ventilation, fume extraction, and PPE, basically giving welders clean air and less reason to glow in the dark.

6.31 How does plasma arc welding (PAW) compare to GTAW welding?

PAW and GTAW appear similar at a glance. Each uses a non-consumable tungsten electrode and an inert gas shield, where any electrode wear is incidental rather than intentional. Yet their personalities split fast. GTAW is the steady craftsperson; PAW is the tightly focused energy cannon.

GTAW welding creates an arc directly between the electrode and the workpiece. It's precise, clean, and perfect for delicate or thin materials, but it's also slow and manual.

PAW, on the other hand, takes that arc and shoots it through a tiny nozzle, turning it into a focused plasma jet. It's faster, deeper, and way more consistent, ideal for automated welding in aerospace, electronics, or anywhere repeatability matters more than setup speed.

GTAW is simpler, cheaper, and widely used. PAW is pricier and pickier, but when you need tight tolerances and blazing speed, it absolutely brings the heat. *Someting to keep in mind, as with anything to do with business in general, the higher the cost, or niche the process, the more you can charge for it as less people are offering said services.*

6.32 What are the challenges of welding titanium or exotic alloys?

Welding titanium and exotic alloys like Inconel, Hastelloy, or Monel isn't for the faint of heart. These metals demand five-star treatment, and if you slip, they punish you fast.

Titanium is especially tricky. When hot, it eagerly absorbs oxygen, nitrogen, or hydrogen from the air, which ruins the weld with embrittlement and rainbow discoloration. The only fix is flawless inert gas shielding, not just on the weld pool but trailing as it cools, think of argon as its bodyguard.

Other exotic alloys have their own chalenges. Cool them too fast or use the wrong filler, and they'll crack, segregate, or morph into metallurgical goblins. Controlling heat input, preheat, and interpass temperatures is critical, and some need post-weld treatments just to stay stable.

Often these metals can only be welded in glove boxes, with lasers or electron beams, under conditions closer to a sci-fi lab than a fab shop.

Welders tackling this need strict procedures, advanced certifications, and nerves of steel, because mistakes are extremely costly. But if you can weld them and still pass X-ray inspection, you've unlocked the good life with steady retainers and world travel to exotic locations.

6.33 How do welding codes and standards (like ASME or AWS) guide quality?

Welding codes and standards, like AWS, ASME, CWB, and ISO, are the rulebooks that keep bridges upright and boilers from going boom. They cover everything from materials and joint prep to qualifying welders and inspecting welds.

AWS D1.1, ASME Section IX, CSA W59, and ISO 3834 are basically welding law, defining what makes a weld acceptable and how to prove it. Ignore them, and you're asking for failed inspections, fines, or worse. There are literally endless volumes describing weld specifications, way beyond the scope of this entry level CNC Q&A book.

6.34 How do you maintain and calibrate welding equipment for consistent results?

Maintaining and calibrating welding equipment is like giving your machine a spa day, minus the cucumber water and with a lot more grinding noises. It's essential if you want consistent, high-quality welds and a machine that doesn't die on you mid-job. Regular maintenance starts with cleaning: torches, nozzles, and contact tips need to stay spatter-free to keep the arc stable and predictable. If you let them gunk up, your weld will start acting like it has a personality disorder.

Cables, hoses, and connectors also need a close inspection, a frayed cable or a leaky hose might not seem like a big deal until your arc starts sputtering like a lawnmower in molasses or your shielding gas takes a detour to nowhere.

Speaking of gas, checking your flow meters and regulators ensures your shielding gas is doing its job: protecting your weld from atmospheric sabotage. Without enough gas coverage, you're looking at porosity, oxidation, and a weld that basically throws up on itself. And don't forget the consumables: tips, liners, and electrodes wear down over time. If you're welding with stuff that's half-used or crusty, expect inconsistent starts, arc wander, and quality that screams "I stopped caring halfway through."

Then comes calibration, which is just a fancy word for "making sure your machine isn't lying through its teeth." Your welder might say it's putting out 120 amps, but if you're not checking it with calibrated test gear, like voltmeters, ammeters, or wire feed speed testers, you could be way off, especially on tight-tolerance or code-required jobs. In small shops, welders often adjust instinctively to less-than-ideal parameters, but with high-end customers who require strict WPS compliance, that flexibility isn't an option.

Calibration should be done periodically, especially when you're working in certified environments, switching processes, or suspecting something's off. And yes, you should write it down. Documentation isn't just for show.

6.35 Welding: Top 10 Takeaways

Joint prep impacts weld strength: *Properly prepared joints improve penetration, reduce inclusions or lack of fusion, and ensure stronger, more reliable bonds by allowing full fusion with the base metal.*

Preheating helps thick sections: *Heating the base metal before welding minimizes thermal shock, reduces cracking risk, and improves fusion in thick or high-carbon steels.*

Post-weld heat treatment relieves stress: *Controlled heating and cooling reduce residual stresses, temper hard zones, and enhance toughness and dimensional stability.*

Shielding gas matters: *Gas type and flow rate affect penetration, spatter, bead shape, and arc stability. Match the gas to the material, position, and process.*

Filler metal selection is critical: *Choosing compatible filler ensures metallurgical bonding, prevents cracking, and maintains required strength and corrosion resistance.*

Porosity signals contamination: *Gas pockets from oil, rust, moisture, or poor shielding weaken welds. Clean surfaces and maintain gas coverage to prevent them.*

Crater cracking can be prevented: *Abrupt stops cause shrinkage cracks. Backfilling or ramp-down techniques help avoid defects at the weld's end.*

Automated systems boost productivity: *Welding automation improves consistency, reduces fatigue, increases output, and lowers rework in high-volume applications.*

Distortion control is essential: *Uneven heat causes warping. Balanced weld sequences, fixturing, and heat control help maintain part alignment.*

Welding codes guide quality: *Standards like AWS D1.1, ASME Section IX, and CSA W59 (Canada) define material, procedure, qualification, and inspection requirements to ensure structural integrity, safety, and code compliance.*

❸ 6.36 Section Quiz

This quiz is like striking an arc, get it wrong, and things can go sideways fast. Melt the facts, not your chances!

1. Why is joint preparation critical for strong welds?
    - A) It ensures proper fit-up, and penetration
    - B) It speeds up welding
    - C) It reduces electricity use
    - D) It improves arc light visibility

2. What is the purpose of preheating before welding?
    - A) Reduce cooling rate and prevent cracking
    - B) Prevent oxidation
    - C) Improve surface color
    - D) Save filler metal

3. What does post-weld heat treatment (PWHT) do?
    - A) Increases residual stress
    - B) Hardens the weld
    - C) Relieves stresses and improves toughness
    - D) Increases weld speed

4. How can you identify undercutting in a weld?
    - A) A groove at the weld toe
    - B) Excess weld metal
    - C) Smooth, shiny surface
    - D) Increased spatter

5. What is interpass temperature control?
    - A) Maintaining electrode temperature
    - B) Cooling the base metal rapidly
    - C) Adjusting amperage after each pass
    - D) Keeping the joint within a specific temperature range between weld passes

6. What's a key difference between AC and DC?
    - A) AC has constant polarity; DC reverses polarity
    - B) DC is only for aluminum
    - C) DC provides smoother arcs; AC helps with magnetic materials and oxide cleaning
    - D) AC is used for thin materials only

7. How do you select filler metal for dissimilar metal welding?

A) Match the stronger metal
B) Match color only
C) Use any available filler
D) Match the weaker or more crack-sensitive metal

8. What is porosity in a weld?

A) Tiny gas pockets trapped in the solidified weld
B) Cracks along the weld bead
C) Excess flux residue
D) A type of electrode coating

9. How do you prevent crater cracking?

A) Increase travel speed
B) Backfill the crater and taper off current
C) Use higher amperage
D) Shorten arc length

10. What is an advantage of pulse GMAW welding?

A) Reduced penetration
B) Lower spatter, controlled heat input, and better out-of-position welding
C) Slower weld speeds
D) Less need for shielding gas

11. How does shielding gas composition affect welds?

A) It only changes arc color
B) It changes electrode polarity
C) It affects penetration, arc stability, and bead profile
D) It reduces slag formation

12. What's a key safety concern when welding galvanized?

A) Electric shock
B) Magnetic arc blow
C) Metal fume fever from zinc fumes
D) Hydrogen cracking

13. What determines weld size in a structural application?

A) Machine size
B) Required throat thickness to carry load
C) Filler metal price
D) Number of weld passes

14. What does visual weld inspection check for?
   A) Underfill, cracks, porosity, and bead uniformity
   B) Cooling rate
   C) Base metal thickness
   D) Flux composition

15. How do automated welding systems improve productivity?
   A) Increase manual labor
   B) Eliminate need for inspection
   C) Reduce power supply needs
   D) Provide consistent weld quality and speed

16. How does laser welding compare to arc welding?
   A) Laser welding uses heat; arc welding uses pressure
   B) Laser welding doesn't require shielding gas
   C) Arc welding is only used for aluminum
   D) Laser welding delivers focused, high-energy beams for precision; arc welding is more versatile

17. What is hydrogen-induced cracking?
   A) Cracking caused by over-welding
   B) Cracking caused by hydrogen trapped in the weld
   C) Cracking due to excessive amperage
   D) Cracking from poor shielding gas flow

18. How does welding affect metal microstructure?
   A) It can alter grain size, hardness, and toughness in the heat-affected zone
   B) It leaves it unchanged
   C) It improves ductility
   D) It eliminates impurities

19. What is the purpose of a bend test?
   A) Evaluate weld ductility and detect internal defects
   B) Check hardness
   C) Measure weld length
   D) Determine electrode quality

20. Why use a backing bar in welding?
   A) Improve weld bead color
   B) Reduce machine vibrations
   C) Increase welding current
   D) Support molten weld pool and ensure full penetration

Quality weldments can be produced without specialized welding tables.

This custom jig ensures accurate rail placement.

✖ 6.37 Answer Key and Explanations

This answer key hits harder than arc blow on a tight root pass, but don't worry, the explanations are smoother than a GTAW weld on clean stainless.

1A – It ensures proper fit-up, cleanliness, and penetration, which are all essential for achieving a sound weld with proper fusion and minimal defects.

2A – It reduces the cooling rate and prevents cracking by minimizing thermal stresses, especially in thick or high-carbon steels.

3C – It relieves residual stresses and improves toughness and ductility in the weld and heat-affected zones, increasing structural integrity.

4A – Undercutting appears as a groove or depression along the weld toe, weakening the joint and potentially leading to cracking.

5D – It refers to keeping the joint within a specified temperature range between weld passes to avoid defects like cracking or lack of fusion.

6C – DC provides a smoother, more stable arc, while AC helps counteract arc blow and is useful for oxide cleaning on materials like aluminum.

7D – You match the filler to the more crack-sensitive or weaker metal to reduce the risk of weld failure due to brittleness or incompatibility.

8A – Porosity is caused by gas entrapment in the molten weld pool that doesn't escape before solidification, weakening the weld.

9B – Crater cracking is avoided by backfilling the crater and tapering off the current to ensure a gradual cooling and complete weld pool solidification.

10B – Pulse GMAW welding offers better control of heat input, reduces spatter, and is ideal for out-of-position or thin material applications.

11C – Shielding gas composition affects penetration depth, arc stability, bead shape, and even spatter levels, depending on the gas mixture.

12C – Galvanized steel produces toxic zinc oxide fumes when welded, which can cause metal fume fever if inhaled without proper ventilation or PPE.

13B – Weld size in structural applications is determined by the required throat thickness, which ensures the weld can safely carry design loads.

14A – Visual inspection looks for surface-level discontinuities such as underfill, cracks, porosity, and irregular bead shapes that may indicate deeper problems.

15D – Automated systems deliver consistent quality, speed, and repeatability while reducing human error, increasing overall productivity.

16D – Laser welding uses highly focused energy for precision and low heat input, whereas arc welding is more adaptable to various positions and materials.

17B – Hydrogen-induced cracking occurs when hydrogen atoms become trapped in the weld metal, especially under tensile stress and cold conditions.

18A – Welding can significantly alter microstructure in the heat-affected zone, affecting properties like grain size, hardness, and overall toughness.

19A – A bend test checks for weld ductility and internal defects like lack of fusion or cracks, ensuring the weld will perform under stress.

20D – A backing bar supports the molten weld pool from behind, ensuring complete joint penetration and preventing weld metal from falling through.

Keep your work laser perfect with acrylic offset shims.

Precision plasma cut base plates integrated into custom railing.

Section 7
Making It All Work Together
Harnessing CNC Tools for Total Freedom

Imagine you walk into a shop that has CNC lasers, a CNC plasma cutter, a CNC 3D printer, and welding gear with lots of woodworking equipment, you're basically stepping into a digital fabrication playground where metal, plastic, composites, and creativity collide. Welcome to my shop!

This isn't just a shop; it's the Avengers-level team-up of manufacturing tech. Each machine has its specialty, but when you combine them, the synergies are nothing short of legendary. In my previous book, I went full Avengers with the teaching style, an unorthodox approach to learning that I hope you enjoyed.

Let's break it down. The CNC laser gives you pinpoint precision on thin materials, perfect for crisp cuts and delicate engraving. Want to cut intricate acrylic panels, fine metal brackets, or custom signage? Laser's your best friend. But hit a thicker steel plate and you'll start sweating. Which is where the CNC plasma storms in, roaring with sparks, blasting through thick metals like a hot knife through cotton. Plasma doesn't care about finesse; it's here for straight-up production speed and power.

But wait, you've got some ultra-thick exotic alloy or delicate glass or stone that neither laser nor plasma can handle? Enter the CNC waterjet, casually cutting anything you throw under its high-pressure stream, no heat-affected zones, no material limits, just raw abrasive precision.

Now, what if you don't want to cut something, you want to create it from thin air? That's where the CNC 3D printer swaggers in. Layer by layer, it builds parts, tools, prototypes, or even finished components that you can then combine with laser-cut, plasma-cut, or waterjet-cut parts. Suddenly, you're mixing subtractive and additive manufacturing like a fabrication wizard, pulling ideas from the digital world into physical form.

7.1 Joining It All Together

All the cutting, printing, shaping, and creating only come together through the magic of joining. For metals, that usually means welding, while for other materials there's a whole library of adhesives to choose from, though their bonding results can be hit or miss compred to welding.

Welding is the ultimate bonding (casing) force. Molten metal that fuses your laser-cut brackets to your waterjet-cut base, or your plasma-sliced structural frames to form a rugged skeleton. Meanwhile, 3D-printed parts, especially those made from plastic or resin, typically don't get welded, they're bolted, bonded, or slotted into place.

Without these joining methods, you'd just have a collection of beautifully crafted parts sitting awkwardly next to each other, like strangers at a party. But with them? You get complete, functional assemblies that are more than the sum of their parts.

7.2 Why Industry loves CNC Synergies

These combined tools offer unmatched flexibility. Instead of being limited by one process, companies can 3D print for prototyping, scale with plasma cutting, tackle tough materials with a waterjet, add precision with lasers, and finish strong with skilled welding.

It shortens lead times, reduces costs, improves product quality, and opens the door to custom solutions no single process could handle alone.

So when you walk into a shop with all these tools (and someone who knows how to use them), you're not just looking at machines, you're looking at a super-powered fabrication arsenal. It's the ultimate mashup of fire, water, light, molten metal, and digital magic, all working together to build the future, one spark, one slice, and one layer at a time.

This is what makes *CNCROi.com* unique, where different capabilities work in synergy to deliver optimal results.

7.3 Hybrid Workflow Safety Myths & Mistakes

The dream shop: a CNC laser, router, plasma cutter, waterjet, and 3D printer all under one roof, humming together like a mechanical symphony. You, the proud owner, stand at the center, arms crossed, surveying your empire of digital fabrication glory. Nothing can stop you now, you've assembled the Avengers of the CNC world!

Except... there's just one tiny problem: combining all these machines also multiplies your opportunities for catastrophic safety mistakes. Welcome to the world of fabrication synergy, where more machines mean more ways to injure yourself, or break expensive toys.

Myth 1: "I can run all the machines at once, it's efficient!"

The legendary multitasker. You hit start on the laser, queue up the waterjet, load a 3D print, fire up the plasma, and spin the router, all at once, baby!

Except now you're racing between stations like a decapitated chicken, trying to catch the router before it eats through your clamps, the laser before it burns the edges, the plasma before it slams into a lifted part, and the 3D printer before it turns a project into spaghetti. Efficiency? Nope. Welcome to the world's noisiest stress test. It's quite rare that I run more than one machine at once.

Myth 2: "I only need one type of PPE to covers everything!"

Sure, just throw on your welding helmet and call it a day, right? Except... the laser wants safety goggles, the router demands ear protection, the plasma cutter spits molten slag at your feet, and the waterjet sprays mist (and high-pressure dangers).

There's no "one-size-fits-all" PPE in a multi-process shop, unless you're planning to walk around in a hazmat suit holding a shield. Match your protective gear to the machine, or you'll end up the wrong kind of shop legend. It might look cool, but a welding helmet does nothing to protect you from the fumes released when ABS is printed.

Myth 3: "One ventilation system to rule them all."

Wrong! Laser fumes? Need a filtered exhaust. Plasma and welding? Need fume extraction. Router dust? Needs dust collection, not smoke filters. Waterjet? Needs splash control, not air scrubbers. And the 3D printer? Needs good room ventilation to avoid that ABS smell on your clothes.

Mixing these systems is like trying to drain a swimming pool with a straw. If you don't match the system to the source, you'll end up choking on a stew of dust, vapor, smoke, and plastic funk.

Myth 4: "My power grid can handle everything running at once."

Ever tried plugging a laser, plasma, router, waterjet, and heater into the same circuit? You'll get a light show when every breaker in the shop trips.

Technically, my landlord is right, I have plenty of power going into my shop, in reality, I still run into power issues and I have solid 3-phase transformers in there!

High-powered machines don't just need space, they need dedicated circuits, clean power, and surge protection. Skip this, and you'll spend half your shop time flipping breakers and explaining to the utility company why your meter suddenly looked like it was powering a small city.

Myth 5: "I can treat toolpaths and materials the same across machines."

The router loves wood, but the laser will set it on fire. The waterjet cuts anything, but the plasma prefers conductive metals. The 3D printer builds up, the laser cuts down, the plasma blasts through, and the router carves away.

Treating materials or toolpaths interchangeably is a recipe for ruined projects and unhappy customers.

Know what each machine wants, or watch your beautifully planned synergy become a synchronized disaster.

Myth 6: "I only need one maintenance schedule."

Wrong. That's like thinking all pets eat the same food, good luck giving kibble to a pet shark.

Every machine has its own quirks. The 3D printer needs nozzle cleaning and bed leveling. The CNC router wants spindle checks and collet tightening.

Plasma cutters chew through electrodes and tips fast, while the waterjet devours abrasive and wears out nozzles and seals. And the laser? It insists on clean lenses and aligned mirrors to stay precise. Don't forget different types of lube for each machine!

Try cramming all that into a single lazy 'maintenance day' and you'll end up with a lineup of underperforming machines. Sure, you can keep putting it off almost indefinitely, but like a cavity, the longer you wait, the more it costs when you finally end up at the dentist. And the dentists of the CNC world are a lot more expensive!

Reuse cutout holes from one job for the next project.

Myth 7: "Noise levels aren't a big deal, it's just background shop sounds."

The router screams, the plasma crackles, the waterjet roars, and even the laser makes its little chirpy hum. Now, pile them together in one room, and suddenly your ears are under full assault. Don't forget dust systems, air compressors and other auxiliary equipment like vacuum tables as well.

Without proper hearing protection, you'll finish the day with your head ringing like a bell. And don't forget: customers or visitors wandering in aren't expecting an audio blast zone. Manage the noise or invest in earplugs, or accept that your new nickname will be "What Did You Say?"

Myth 8: "Fire risk isn't multiplied, I only have one shop, not five!"

A plasma torch can throw sparks; a laser can ignite thin materials; routers can overheat dust; 3D printers can short-circuit; waterjets, okay, they're less fiery (but don't tempt fate). Multiply your machines, and you multiply your fire points. You'd be surprised how fast things can happen.

Don't think of it as one shop, think of it as endless opportunities to set something ablaze. Fire extinguishers, alarms, and a clear emergency plan aren't optional when your shop looks like an A-Team's showroom.

Myth 9: "I can just focus on one machine, the others will run fine unattended."

The unattended router will have a dust explosion. The laser will flame out on the edge. The plasma will wander off-path, dragging your sheet with it. The 3D printer will clog. The waterjet? It's probably fine, until it isn't.

Multi-machine shops demand constant attention, smart setups, and efficient workflow. Treating it like a "set-and-forget" factory is how you end up chasing simultaneous emergencies like a one-person pit crew. Lights out manufacturing does exist, just most likely not for you.

Myth 10: "The synergy means I can do anything now!"

Having all these tools expands your capabilities, but it also expands your failure points. Just because you can combine 3D-printed parts, laser-cut details, plasma-cut brackets, and waterjet-cut fine inlays doesn't mean it'll all work perfectly together without testing, design adjustments, or sometimes, total reworks.

Synergy gives you power, but without careful planning, you'll create Frankenstein parts: overengineered and probably destined for the scrap bin.

7.4 More Machines, More Smarts

Owning a multi-process shop like *CNCROi.com* is awesome, a total fabrication flex. With lasers, routers, plasma cutters, waterjets, and 3D printers all under one roof, you can tackle jobs other shops only dream about. You're not just building parts; you're building possibility. But with great tools comes an even greater circus of safety risks, each one just waiting to catch you off guard.

Every machine brings its own hazards: lasers can blind, plasma can burn, routers can kick back, and waterjets? Throw a jet of water on your electrical panel. Even the seemingly innocent 3D printer can quietly fill your workspace with toxic fumes if ventilation is ignored. It's not about fear, it's about respect.

Respect each machine. Match the right tool to the right task. Invest in proper safety setups: fume extraction, grounding, guards, fire extinguishers, and emergency shutoffs. Train like it matters, because it does. And most importantly, don't assume that more machines make you invincible. They just make you the proud owner of a very powerful, very complex dance floor, where every move counts and one misstep can cost you time, money, or worse.

So slap on those goggles, earmuffs, gloves, and respirators. Double-check your materials, inspect your connections, clear your work area, and power up your machines for the sympony that's awaiting your ear plugs.

☐ Clean-up the Shop, It's Q&A Time!

7.5 How do you decide which CNC process is best for each stage of a hybrid project?

Picking the right CNC process for each stage of a hybrid project is like casting a movie where every machine has a role, and you don't want your lead actor showing up in the wrong scene. You've got to juggle material type, precision needs, speed, cost, and what each machine can actually handle without throwing a mechanical tantrum.

Say you need to hog out a bunch of pieces fast, that's your CNC router or plasma cutter's time to shine. They're the brute-force muscle, not the finesse artists. But when it comes to delicate engraving or crisp surface marking, you'll want to call in the laser, the quiet, precise type that doesn't break a sweat when dancing on microns.

Material compatibility is a whole other game. Try running wood through a plasma cutter and you'll end up with grounding issues, not a part. Routers love wood, plasma cutters eat steel for breakfast, and lasers? Lasers are picky but fabulous with plastics and thin metals.

Then there's the question of detail. Need tight tolerances or perfectly fitting parts? Waterjet or laser. Don't care if it's off by half a millimeter? Plasma's your pal.

And let's not forget timing, where some steps, like 3D printing, might come early in the process, building up parts before anything gets cut due to long production times.

Welding and finishing are the final steps that bring all your parts together into a complete assembly. Be sure to silicone those tapped holes before galvanizing or you will create a lot of extra work for yourself.

The key is to think like a symphony conductor: break the job into stages, pass the baton to the right machine at the right moment, and never let the woodwinds play the drums or the cello take on the xylophone.

7.6 What are the advantages of combining laser engraving with CNC-routed components?

Mixing laser engraving with CNC routing is like pairing muscle with style, think lumberjack in a tailored suit. You get the beefy, structural chops of routed parts and the fine-detail flair only a laser can deliver.

Routers are the heavy lifters. They're great at shaping big, thick, or complex parts out of wood, plastic, or composite. Need curves, pockets, joinery, or beefy panels? The router's your workhorse. But ask it to engrave a logo or etch fine lettering and it'll look at you like, "Buddy, I carve depth, not beauty." You always have that CNC bit diameter limitation.

That's where the laser swoops in: precise, elegant, and barely touching the surface. It adds logos, patterns, serial numbers, or custom art without breaking a sweat (or your parts). No clamping gymnastics. No messy tool changes (get an ATC). Just fast, clean, surface-level goodness.

This tag-team approach is a favorite in signage, furniture, packaging, and architectural panels, anywhere you want strong parts and sharp looks. For small shops, it's a game-changer. You boost the value of your products without tanking your production time. The router builds the body, the laser adds the soul.

7.7 How can plasma-cut steel parts be integrated into 3D-printed assemblies?

Mixing plasma-cut steel with 3D-printed plastic is like building a cyborg; strong on the outside, custom on the inside. But it's not plug-and-play. Plasma parts bring the muscle, but only if your 3D prints are designed to welcome them with perfectly sized slots, pockets, or mount points (don't forget to account for kerf and taper).

Deburr the steel, clean it up, and skip the glue, using screws, bolts, or rivets are your best bet unless you want your robot arm flinging itself apart mid-demo. This hybrid combo is ideal for robotics, machines, or artsy builds where strength meets flexibility.

7.8 What are the challenges of welding CNC-cut parts from different machines?

Welding CNC-cut parts from different machines sounds like a solid plan, until your parts start acting like they're from rival high schools.

The laser kid's all polished and precise, the plasma jock shows up loud and scorched, and the waterjet art student? Clean but definitely doing its own thing. Spoiler alert: not all cuts play nicely together.

That sleek laser-cut part? Yeah, it might be hiding a thin oxide layer that'll totally ghost your weld. The plasma-cut piece shows up with hardened edges and some burnt-on baggage, while the waterjet-cut part, though heat-free, may be just rough enough to make you question your surface prep choices. Each one has its quirks, and unfortunately, your welder isn't a therapist.

Even when the materials look the same, tiny differences in alloy composition or sneaky coatings can throw off your weld quality. Toss in mismatched tolerances, because, surprise! Each machine speaks its own dialect of precision, and suddenly your perfectly engineered assembly requires a crowbar, a belt sander, and possibly a minor exorcism to fit together unless your tolerances are within 6 mm (1/4").

And don't sleep on residual stress: laser and plasma cuts might already be carrying a whole lot of internal tension just waiting to warp your dream project mid-weld.

So, what's the fix? Treat every part like it's got history. Give it a proper cleaning, smooth the edges, double-check fitment, and maybe throw in some stress relief if it's been through a particularly hot breakup (read: plasma cutter). Don't assume your parts will just vibe because they're cut to spec, welding across machine types is less plug-and-play and more diplomacy-meets-grit.

Get it right, though, and you'll end up with a rock-solid hybrid build, a beautiful alliance of metal personalities that, somehow, makes it all work. Even the plasma guy.

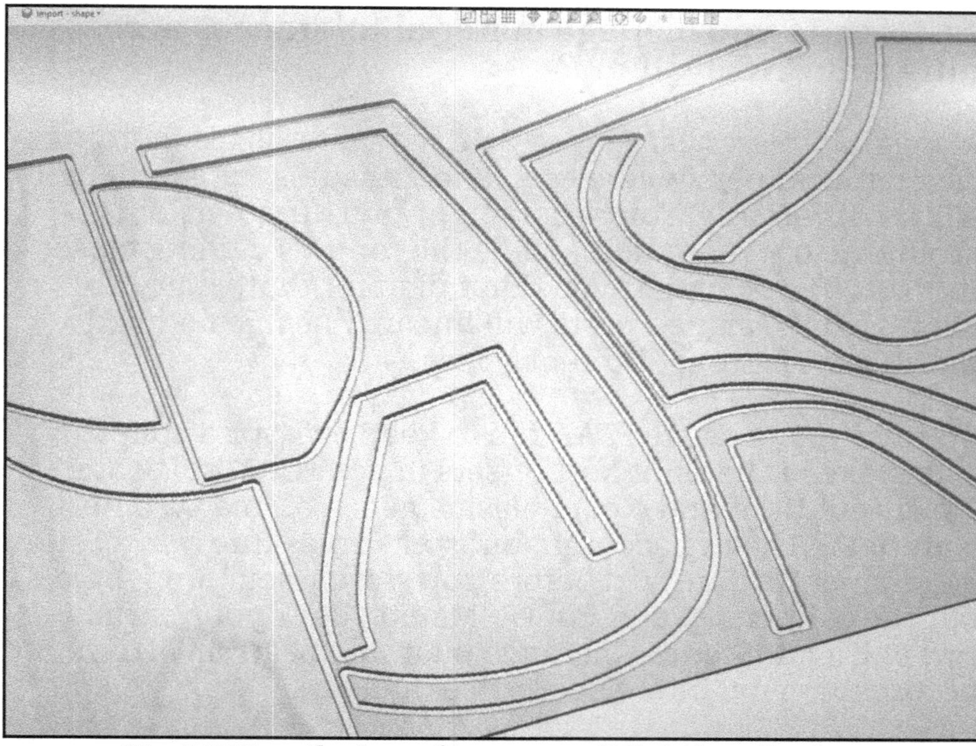

Use MDF as the base for your wall lighting display.

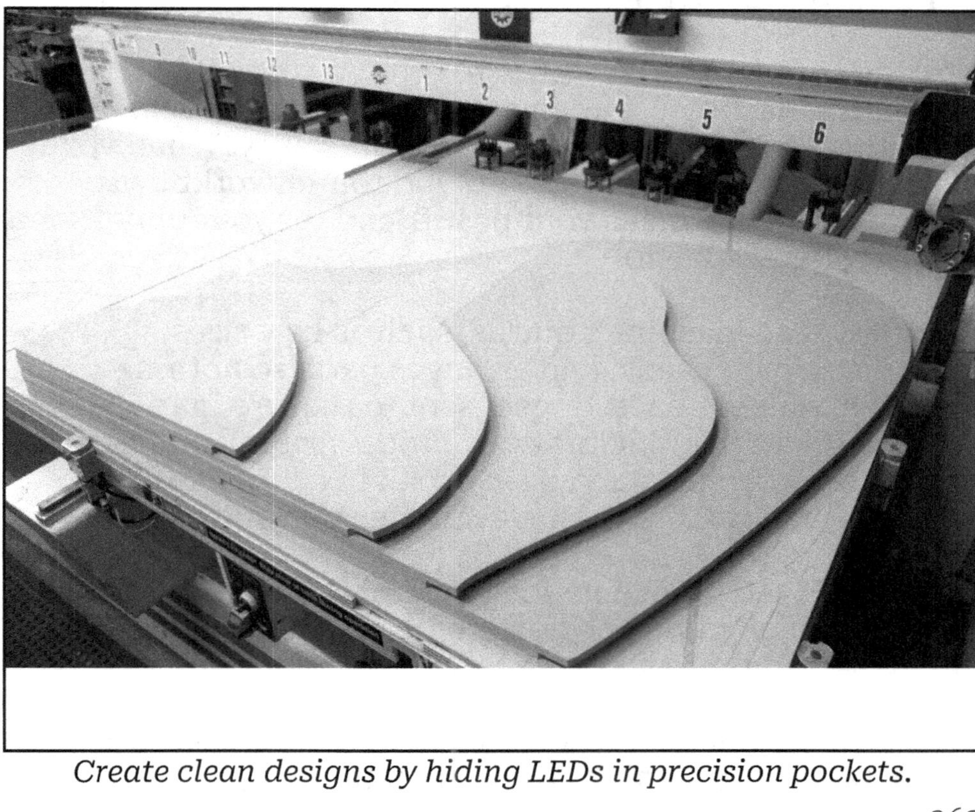

Create clean designs by hiding LEDs in precision pockets.

7.9 How do you manage material tolerances across different CNC systems?

Managing tolerances across CNC machines is like herding cats with calipers where each one measures "close enough" a little differently. Your laser might pull off precision around ±0.004" (±0.1 mm), while the router's chilling closer to ±0.012" (±0.3 mm). If you don't plan for that, your beautifully designed parts will line up like mismatched puzzle pieces from different boxes.

Start with one master CAD file to keep everyone honest, then bake in the right kerf offsets in your CAM software. Lasers cut tight and clean, plasma gets a bit melty, and waterjets? They're accurate but sometimes throw in a little edge taper for fun. And before you start assembling, break out the calipers, gauges, or CMM and check your parts. A few extra minutes of tolerance planning beats an afternoon of forcing metal into place with a mallet.

7.10 What role does nesting software play when combining multiple machine workflows?

Nesting software is like Tetris for grown-ups with expensive machines, it crams your parts onto sheets in ways that save material, time, and your sanity. When you're juggling laser, plasma, waterjet, and router workflows, smart nesting becomes your best friend (or your only friend when deadlines loom).

It knows each machine's quirks, such as kerf sizes, tolerances, cut speeds, and arranges parts so nothing overlaps, wastes space, or gets sent to the wrong tool. It can group by material, thickness, or process, crank out machine-ready code, and even sneak in tricks like common-line cutting to shave minutes off production.

Advanced systems can juggle multi-process projects too, flagging which parts need engraving, tapping, or bending, and which ones are just here to be sliced and sent.

Skip nesting, and you're signing up for scrap piles, idle machines, and long nights of wasted time.

7.11 How do you streamline file preparation for multi-machine fabrication?

Want to avoid chaos in your CNC shop? Start with a centralized CAD model, your single source of truth, your design gospel. I know I just said that on the previous page but this is VITAL! From that holy file, you'll branch off into CAM software tailored to each machine: laser, router, plasma, and 3D printer. All CNCs want their own toolpaths, offsets, and machine codes like picky eaters at a buffet.

Keep your file names consistent (no more final_final_THIS_ONE_v3.dxf), stick to universal formats like DXF or PDF so your machines actually talk to each other, and whatever you do, don't bake machine settings into the base CAD. That's like drawing permanent mustaches on your blueprints, you'll regret it later.

Oh, and decide who's prepping, checking, and sending files where, because "I thought you had it" is not a workflow. Streamlined file prep means fewer headaches, less rework, and machines that start cutting instead of arguing with your laptop.

7.12 How can 3D printing be used to prototype parts for plasma or laser cutting?

3D printing is like a dress rehearsal for your metal parts. Before you start slicing steel or frying aluminum with lasers, print it in plastic first and see if your design actually works in the real world, generally at a fraction of the scale. You can test fit, check hole patterns, confirm interlocking features, and avoid that "oops" moment that comes after burning through a $950 sheet of 316 stainless.

It's also a great way to show off your design to clients or teammates without waving around a CAD model and hoping they understand what "fillet radius" means. For complex builds, you can even prototype the assembly process, spot clearance issues, or make custom jigs and fixtures before touching metal. 3D printing gives you a cheap, fast feedback loop that makes your final cuts smarter, cleaner, and way cheaper.

7.13 What are the benefits of combining waterjet and plasma cutting on thick metal projects?

Combining waterjet and plasma cutting is like pairing a chainsaw with a scalpel; rough cuts where speed matters, precision where it counts. Same idea as the router vs laser.

Plasma blasts through thick metal fast and cheap, but leaves behind a heat-affected zone and a kerf wide enough to park a truck. Waterjet, on the other hand, is slow, pricey, and clean enough to cut wedding rings out of titanium.

The move? Use plasma for the big, messy work, then let the waterjet handle the fine details, such as bolt holes, tight fits, or anything your engineer would cry about if scorched. This hybrid tag-team speeds up production, saves money, and cuts down on post-processing.

If you want to get into the CNC business, make sure your setup includes both a fast, rough-cut machine and a slow, high-precision one.

Custom octagon window frames with easy slide in waterjet cut glass.

7.14 How does post-processing differ when integrating parts from laser, plasma, and router workflows?

Post-processing parts from different CNC machines is like cleaning up after three very different roommates. The laser guy's pretty neat, just a little oxidation or discoloration, maybe a quick wipe-down.

Plasma? Total slob. Expect dross, heat-affected zones, and some serious grinding therapy. And the router? Usually covered in fuzz, splinters, or burrs, especially if this wood chipper been chewing through wood or composites.

If you're bringing all these parts together into one assembly, don't leave cleanup to the last minute. Match your post-processing to the goal, whether you need tight fits, bondable surfaces, or a polished, professional finish. A little extra prep now saves you from discovering mid-assembly that nothing will fit together properly.

7.15 How do you balance precision and speed across machines to meet production deadlines?

Balancing precision and speed is like choosing between a race car and a scalpel; both are great, just not for the same job. Laser cutting gives you sharp detail but slows down on thick metal. Plasma's faster, but its edge quality can be more "functional" than fancy. Routers? Awesome for wood and plastics, but don't expect hair-splitting accuracy.

If you're on a deadline, play to each machine's strengths. Let plasma blast out rough shapes, save the laser for the high-precision bits, and run as many machines in parallel as possible, don't bottleneck your whole project behind one machine. There are of course other CNCs out there with other optimal settings depending on the material or process.

Smart nesting, tighter toolpaths, and knowing which parts actually need perfection will save you time and headaches.

Because sure, accuracy matters, but not if the part shows up too late to be useful or worse, out of tolerances.

7.16 What safety considerations arise when moving parts between CNC and welding stations?

Moving parts from the CNC table to the welding station might seem routine, until someone grabs a still-sizzling steel plate. Freshly cut metal can be razor-sharp (even coming from a plasma), and most likely, still scorching hot.

Then there's the bonus danger: parts that look innocent but are coated in oils or films that turn into toxic fog the moment the welder sparks up. Plasma is hot but only where it cuts. Add in the fun of juggling multiple alloys without proper labeling, and suddenly you've created a chemistry experiment no one signed up for.

The fix? Cool stuff down, clean it up, glove up, label everything, and train your crew like safety isn't optional, because it's not. This will also help with heat distortion.

At *CNCROi.com*, I generally work with either stainless or mild steel to keep things simple. It's easy to visually identify even is some aluminium is in the mix.

7.17 How can robotic automation tie together multiple CNC processes on one production line?

Robotic automation is like the shop's overachieving intern, it connects your CNC processes, and somehow always knows exactly where to put things. Picture a robotic arm grabbing parts fresh off the laser cutter, tossing them (gently) to the press brake, then sliding them into the welding cell, all without anyone lifting a finger or a finger getting sliced off. This capability is coming to small shops!

Robots boost efficiency by skipping breaks, improve precision by never eyeballing a part, and make your shop safer by keeping humans away from hot, sharp, or heavy stuff. Plus, when tied into your Manufacturing Execution System (MES), they track jobs better than most people do.

Sure, setup isn't cheap and programming takes some brainpower, but once it's running, robotic integration turns your workflow into a high-speed, low-error production.

7.18 What software tools best manage hybrid fabrication workflows?

Managing a hybrid fab shop without the right software is like trying to conduct an orchestra with oven mitts. You need tools that connect design, CAM, scheduling, and machine ops without forcing you to retype file names for the fifth time. Something thankfully most small shops don't need to worry about until they get a few employees deep.

Integrated platforms like Fusion 360 or SolidWorks with CAM let you prep toolpaths for multiple machines without jumping between a dozen apps. MES and ERP systems like JobBOSS or ProShop keep your scheduling, materials, and job tracking from descending into spreadsheet chaos. And for cutting workflows, nesting software like SigmaNEST or ProNest makes sure you're not wasting half a sheet of steel every time you hit "Go."

Sometimes a full-stack solution works, but often, a mix of specialized tools plays nicer with your shop's size and sanity. The real win? Systems that talk to each other, keep humans out of the loop where they shouldn't be, and help your CNCs actually get along.

7.19 How do you coordinate finishing techniques when combining metal, wood, and plastic components?

This is a craft-meets-chaos. Wood wants stain, metal wants powder coat, plastic just wants to be left alone, and somehow, they all need to look like they belong together.

The secret? Don't treat finishing like an afterthought. Pre-finish when possible, as it beats masking around weird geometry later. Choose coatings and adhesives that actually stick to all your materials, and work in order: sand the rough stuff, clean the greasy stuff, and don't let one material's mess ruin another's finish. I find enamel paint works with just about everything.

Your craftsman's eye sees details most people overlook, so reserve your perfectionism for the features that really make a difference.

7.20 How can additive (3D printing) and subtractive (routing, cutting) methods complement each other?

Additive and subtractive manufacturing are the odd couple of the shop floor. One builds wild, complex shapes out of thin air, while the other shows up with precision tools and cleans up the chaos. 3D printing tackles the impossible: internal channels, lattice structures, and organic forms no end mill could touch. Then CNC milling or routing steps in to flatten surfaces, drill perfect holes, and dial in the tolerances like a pro.

This hybrid tag-team shines in aerospace, medical, mold-making, anywhere you want lightweight complexity and engineering-grade accuracy. Print a near-net shape to save material and machine time, then let your CNC finish the critical dimensions. You can even print jigs, fixtures, or prototypes that support and guide traditional machining.

The result? Faster iterations, less scrap, and parts that push the boundaries of design. Proof that printing and cutting, when combined, are greater than the sum of their parts.

Router cut Corian. Laser engraved. Jon paint fill.

7.21 What are the pitfalls of mismatched kerf, taper, or edge quality across machines?

Combining parts from different CNC machines can feel like trying to fit puzzle pieces from three different types of Lego sets. A laser-cut part might look sharp and elegant with its tiny kerf, while the plasma-cut chunk next to it shows up rough, wide, and a little too "enthusiastic." Waterjets? Precise, but they might sneak in a taper if the material's thick. And CNC routers? They'll round every internal corner like they're allergic to sharp inside angles.

To dodge the chaos, design like you know the machines are out to mess with you. Offset for kerf, plan for taper, and remember: mismatched parts don't magically align in final assembly, unless you like filing metal at 2 AM.

Mixing CNC methods without planning is like throwing a dinner party where every guest brings soup and no one brought bowls. Design smart, and test early.

7.22 How does shop layout affect the efficiency of multi-machine production?

Shop layout is the unsung hero of productivity, or the villain behind every tripping hazard and forklift traffic jam.

In a multi-machine shop, a bad layout means workers dragging heavy parts across the floor like medieval peasants, dodging bottlenecks, and racking up steps like they're training for a marathon they never signed up for.

A smart setup keeps machines arranged like a well-planned buffet: in the order you need them, with no backtracking. Got CNC routers, lasers, and plasma tables? Put them near the material stash, not in opposite corners like estranged relatives. Keep welding and finishing close too, unless you enjoy buying lots of propane for your forklift.

Mark the floor like it's a board game, leave breathing room for storage and staging, and plan like future-you will one day add that shiny new machine.

7.23 How can modular jig systems support cross-machine precision assembly?

Modular jig systems are like the LEGO sets of the workshop world, being infinitely adjustable, and reusable. They keep your oddly matched parts (looking at you, laser-cut steel and 3D-printed ABS) from wandering off during welding, gluing, or bolting like rebellious toddlers.

Instead of building a new jig every time your design team "has a cool idea," you just reconfigure your trusty modular setup. I know, easier said than done. It's like having a Swiss Army fixture that doesn't judge your part tolerances.

Clamp here, align there, drop in a locator pin, and boom, your Frankenstein assembly suddenly fits like it was born that way. Perfect for custom work, or small batches.

7.24 How do you estimate project costs and timelines when combining multiple fabrication methods?

Estimating hybrid projects is part math, part magic. Each process, of which there can be an infinite amount of steps, has its own costs and quirks. Plasma's cheap but messy, laser's clean but pricey, and prints take forever plus post-curing. Welding adds labor and sparks (literal and financial).

To get it right, break it down:

- What machines, how long, in what order?
- Material costs (don't forget scrap).
- Labor rates for both machines and humans.
- Setup and changeover time.
- A buffer, because something will go wrong somehow.

Use spreadsheets, ERP, or quoting software to make sense of it all. Solid estimates win jobs, protect profits, and keep surprises to a minimum.

Like most things that you do enough of, eventually you develop a gut instinct for pricing by going through the steps in your head like a fast-forwarded TV show.

7.25 Hybrid: Top 10 Takeaways

Choose the right process for each stage: *Match the job to the machine that plays to its strengths, laser, plasma, or router, for maximum efficiency and results.*

Hybrid workflows unlock new possibilities: *Combining CNC, welding, and 3D printing expands capabilities beyond any single process.*

Tolerances must be managed across machines: *Because machine precision varies, these tolerances need to be considered when designing hybrid assemblies.*

Nesting software optimizes multi-machine jobs: *Software tools streamline material use and file prep across systems.*

Prototyping with 3D printing saves time: *Print fit-check models or tool jigs before committing to full-scale machining.*

Post-processing needs coordination: *Plan surface treatments and secondary ops for compatibility when combining parts.*

Safety across stations matters: *Moving parts between machines and welding stations introduces cross-process hazards.*

Automation ties it all together: *Robotics and software integrate every CNC stage into one seamless production line, improving speed, flow, and overall efficiency.*

Modular jigs improve precision: *Reusable jigs support cross-machine setups, improving alignment and repeatability.*

Cost and time estimates require full-picture thinking: *Combining methods affects lead times, material use, and final pricing, plan accordingly.*

⊗ 7.26 Section Quiz

This quiz is like a hybrid build, sync it right, it flows; miss, and it jams.

1. How do you choose the right CNC process for each stage?
   - A) Use the fastest machine for all steps
   - B) Match material, precision, and process strengths
   - C) Always start with laser cutting
   - D) Only use additive processes

2. What's an advantage of combining laser engraving with CNC-routed parts?
   - A) Faster cutting speeds
   - B) Reduced need for material prep
   - C) Decorative detail and structural strength together
   - D) Elimination of post-processing

3. How can plasma-cut steel parts be integrated into 3D-printed assemblies?
   - A) Glue without preparation
   - B) Design fit features, then join with fasteners or glue
   - C) Use high-speed sanding
   - D) Only weld them

4. What's tricky about welding parts from different CNCs?
   - A) Different edge conditions and tolerances
   - B) Color mismatch
   - C) Software incompatibility
   - D) Too much dust

5. How do you manage tolerances across different CNCs?
   - A) Ignore small differences
   - B) Only work on thin materials
   - C) Switch machines for every cut
   - D) Use a shared CAD design

6. What does nesting software help optimize?
   - A) Material use and machine efficiency
   - B) Filament drying
   - C) Shield gas mixing
   - D) Electrode sharpening

7. How can you streamline file preparation across CNCs?
   A) Use consistent CAD/CAM workflows
   B) Always work in STL files
   C) Change designs mid-process
   D) Ignore kerf differences

8. How is 3D printing useful before plasma or laser cutting?
   A) It isn't
   B) It creates molds for cutting
   C) It prototypes designs for fit and function
   D) It strengthens metal parts

9. What's the benefit of combining waterjet and plasma?
   A) Higher heat input
   B) Improve edge burn
   C) Eliminate abrasive use
   D) Faster cuts (plasma) and precision (waterjet)

10. Why does post-processing vary between CNC methods?
   A) Each process leaves unique edges and residues.
   B) Post-processing is unnecessary
   C) Only plasma cutting needs cleanup
   D) Laser cuts always need welding

11. How do you balance speed and precision across CNCs?
   A) Always prioritize precision
   B) Match each machine's strengths to project needs
   C) Use the same speed settings for all machines
   D) Only run one machine at a time

12. What's a key safety concern when moving parts between CNC and welding stations?
   A) Color matching
   B) Electrical compatibility
   C) Sharp edges, residual heat, and material ID
   D) Software updates

13. How does robotics boost multi-CNC production?
   A) Eliminate the need for human workers
   B) Provide consistent part handling, reduce downtime, and improve precision
   C) Slow down production
   D) Reduce machine lifespan

14. Best software for managing hybrid workflows?
A) Integrated CAD/CAM, MES, or ERP systems
B) Photo editing software
C) Only slicer programs
D) Welding power supply firmware

15. Why is it important to coordinate finishing techniques?
A) Different materials react differently to sanding, painting, or coating
B) All materials can be finished the same way
C) Only metal needs finishing
D) Finishing is optional

16. How do additive and subtractive methods complement each other?
A) They don't
B) Additive only works for metal
C) Only subtractive adds strength
D) Additive builds; subtractive refines

17. What's a pitfall when combining cuts from different CNCs?
A) Perfect edge alignment
B) Less material waste
C) Increased cutting speed
D) Mismatched kerf, taper, or surface quality

18. How does shop layout affect multi-machine efficiency?
A) Layout has no effect
B) Only layout of 3D printers matters
C) Placing machines randomly improves creativity
D) Smart layout cuts part movement and handle time

19. What do modular jig systems help with?
A) Holding parts precisely during cross-CNC assembly
B) Increasing cutting power
C) Improving filament quality
D) Cleaning metal surfaces

20. How to estimate costs and timelines for hybrid projects?
A) Guess based on past jobs
B) Ignore machine time
C) Break down each process, material, labor, and setup
D) Only calculate material weight

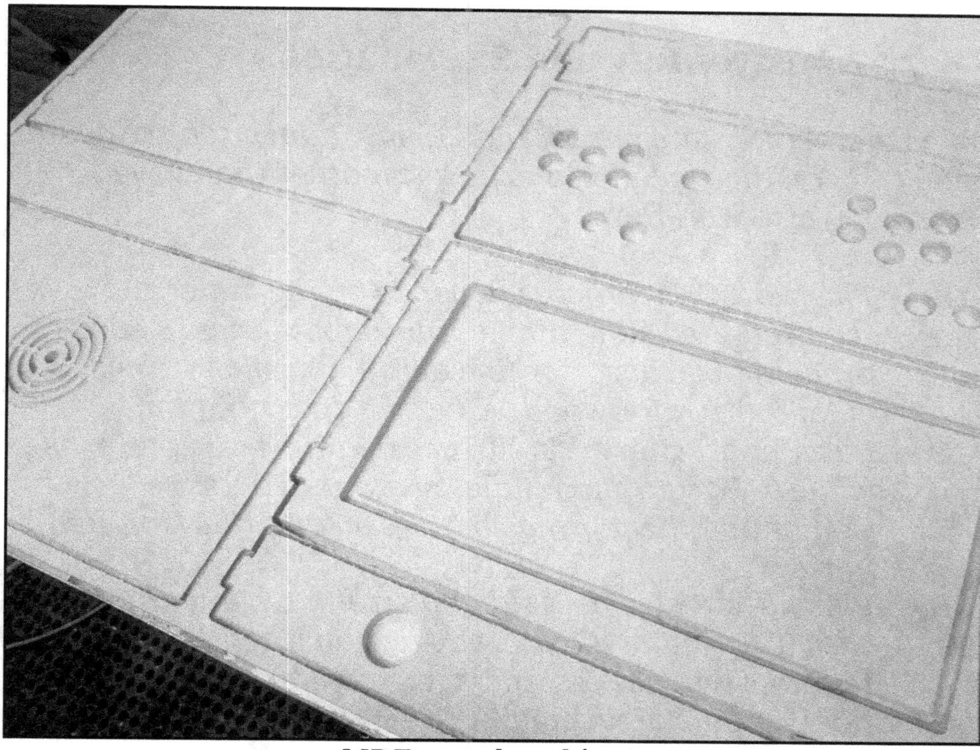

MDF arcade cabinet.

Transform your project by integrating custom electronics.

✘ 7.27 Answer Key and Explanations

This answer key hits harder than a misaligned toolpath, but don't worry, the explanations flow smoother than a well-synced hybrid workflow.

1B – *The right CNC process depends on the material type, required precision, and unique capabilities of each process (e.g., laser for fine detail, plasma for thick cuts, routing for wood). One size does not fit all.*

2C – *Laser engraving adds surface detail (logos, patterns), while CNC routing handles structural cuts. Combining them brings both form and function to the part.*

3B – *Plasma-cut steel is rigid and strong. To integrate it with 3D-printed plastic, you design interlocking features and use fasteners (e.g., bolts) or adhesives, not just glue or welding.*

4A – *Parts cut on different machines often have varying kerfs, edge qualities, or heat-affected zones, which can complicate welding alignment and integrity.*

5D – *Managing tolerances requires a single design source and applying adjustments for each machine's precision limits (like kerf offsets or shrinkage factors).*

6A – *Nesting software arranges parts on a sheet to reduce scrap material and optimize cutting paths, saving both time and cost.*

7A – *Keeping CAD/CAM workflows uniform and using version control systems (like Git or PDM tools) ensures file consistency and reduces errors during transfers.*

8C – *3D printing allows fast prototyping to check form, fit, and function before committing to more costly plasma or laser cutting of metals.*

9D – *Plasma cuts thick metal quickly, while waterjet can refine edges with high precision and no heat distortion, making them a powerful hybrid combo.*

10A – CNC, laser, plasma, and 3D printing leave distinct marks (e.g., burrs, soot, support material). Post-processing must be tailored to the specific residue or finish issue.
11B – A smart workflow matches fast tools (like plasma) for roughing, and precise tools (like milling) for final passes, balancing speed with end quality.
12C – Transferring parts between machines involves safety risks like hot metal, sharp corners, or confusing similar-looking materials. Safety protocols are essential.
13B – Robotic arms or conveyors can move parts between stations, improving repeatability, reducing handling errors, and streamlining throughput.
14A – Managing hybrid workflows across machines is best done with integrated software systems (e.g., Fusion 360, Siemens NX, ERP platforms) for data, scheduling, and tracking.
15A – Wood, metal, and plastic each have unique finishing needs (e.g., melting point, surface adhesion), requiring different approaches for quality and consistency.
16D – Additive manufacturing creates complex internal geometries, while subtractive tools (e.g., milling, turning) are used to fine-tune surfaces and tolerances.
17D – Different machines create different edge profiles. Without proper planning, the parts may not align, requiring rework or extra finishing.
18D – Efficient shop layout means less time wasted moving parts between machines, which improves workflow speed, reduces damage, and improves safety.
19A – Modular jig systems provide repeatable, adjustable setups for accurate alignment during welding, drilling, or routing across different machines.
20C – Accurate hybrid workflow planning involves evaluating each stage's cost (machine time, material usage, labor, and setup), not just total weight or guesswork.

⊗ Wrap-Up Quiz

Let's see if all those late-night cram sessions actually paid off! Write down why that's the right answer too, because "I guessed" won't cut it unless you're playing Jeopardy.

1. What's the purpose of a compression bit on CNC routers?
   A) Increase cutting speed
   B) Reduce tear-out on both top and bottom surfaces
   C) Improve spindle cooling
   D) Avoid tool breakage

2. Why is abrasive particle shape important for waterjets?
   A) Spherical particles increase cutting force
   B) Rounded particles reduce nozzle wear
   C) Particle shape doesn't matter
   D) Sharp particles enhance cutting efficiency

3. How does a dual-extruder 3D printer improve capabilities?
   A) Faster single-material prints
   B) Enables multi-material or multi-color printing
   C) Reduces the need for slicer software
   D) Eliminates the need for supports

4. Why use a downdraft table for CNC plasma cutting?
   A) Increase cutting speed
   B) Provide cooling water
   C) Reduce operator exposure to fumes and particles
   D) Sharpen consumables

5. Why do CO_2 lasers require regular mirror alignment?
   A) To keep the laser beam focused along the entire bed
   B) To increase cooling efficiency
   C) To change the wavelength
   D) To sharpen the beam edges

6. When welding, why is back-purging used on stainless pipes?
   A) Prevent oxidation on the inside of the joint
   B) Improve surface color
   C) Reduce filler wire cost
   D) Increase weld current

7. What is kerf compensation in CNC programming?
   A) Accounting for cut width in the toolpath
   B) Adjusting tool diameter
   C) Increasing spindle RPM
   D) Changing material thickness

8. How does waterjet pump type affect performance?
   A) Direct drive pumps are better for high-volume cutting; intensifier pumps handle higher pressures
   B) Direct drive pumps reduce water use
   C) Pump type doesn't affect performance
   D) Intensifier pumps eliminate abrasive needs

9. What's a benefit of 3D printing with a heated build chamber?
   A) Reduce filament consumption
   B) Improve adhesion and reduce warping
   C) Speed up slicer calculations
   D) Lower printer cost

10. What does arc voltage sensing do in a CNC plasma?
   A) Measures material thickness
   B) Increases cutting gas pressure
   C) Adjusts torch height dynamically
   D) Switches cutting gases

11. Why is laser cutting acrylic known for polished edges?
   A) The material evaporates cleanly under laser heat
   B) Acrylic doesn't need cutting, only engraving
   C) The laser melts and smooths the edges
   D) The air assist acts as a polish

12. What does weld porosity indicate?
   A) Gas entrapment during solidification
   B) Excessive heat
   C) Low arc voltage
   D) Too fast travel speed

13. How can modular jigs improve hybrid fabrication?
   A) Reduce machine power
   B) Allow precise, repeatable positioning
   C) Increase cutting amperage
   D) Eliminate fixturing needs

14. Why is spindle runout a problem in CNC routing?
    A) It improves surface finish
    B) It increases tool sharpness
    C) It reduces motor power
    D) It causes wobble, reducing precision and tool life

15. How does garnet mesh size affect waterjet cutting?
    A) Finer mesh cuts faster
    B) Coarser mesh increases cutting precision
    C) Mesh size only affects water flow
    D) Coarser mesh improves speed but reduce edge quality

16. What's the purpose of a brim or raft in 3D printing?
    A) Improve bed adhesion for the part
    B) Increase nozzle temperature
    C) Reduce filament cost
    D) Enable multi-color printing

17. Why is torch height control critical in plasma cutting?
    A) Prevent consumable wear and ensures arc focus
    B) Increase amperage
    C) Reduce shield gas use
    D) Improve fume extraction

18. How does beam spot size affect laser engraving resolution?
    A) Larger spots increase detail
    B) Spot size has no effect
    C) Smaller spots increase detail
    D) It only matters for cutting, not engraving

19. What's a safety concern when welding aluminum?
    A) Flammable dust
    B) High UV radiation and invisible oxide layer
    C) Low-temperature fumes
    D) Magnetic fields

20. How can software nesting reduce material costs?
    A) By increasing machine weight
    B) By packing parts tightly to minimize scrap
    C) By lowering spindle RPM
    D) By increasing tool changes

21. In CNC routing, what is the effect of using a downcut bit?
   A) Clears chips upward
   B) Increases heat buildup
   C) Pulls fibers downward for cleaner top edges
   D) Improves 3D contouring

22. What's a reason for using a mixing tube in CNC waterjet?
   A) Reduce abrasive cost
   B) Mix abrasive with water jet to create cutting power
   C) Filter out contaminants
   D) Cool the nozzle

23. Why are retraction settings important in 3D printing?
   A) Control filament pull-back to reduce stringing
   B) Improve bed adhesion
   C) Adjust layer height
   D) Change nozzle diameter

24. What cutting condition increases the risk of double arcing in plasma cutting?
   A) Increasing cutting gas flow
   B) Using a worn nozzle or electrode
   C) Using a water table
   D) Decreasing pierce delay

25. What's the benefit of using a fiber laser over a CO_2 laser?
   A) Better for nonmetals
   B) Lower precision
   C) More efficient cutting of reflective metals
   D) Requires more maintenance

26. How does weld bead profile affect mechanical strength?
   A) No effect
   B) Excessive convexity or underfill can weaken the joint
   C) Only surface finish matters
   D) Higher reinforcement always improves strength

27. How can you ensure cross-machine precision when combining CNC and welding?
   A) Skip calibration steps
   B) Increase weld amperage
   C) Use precise fixturing and maintain tolerances
   D) Reduce design complexity

28. What is a sign of under-extrusion in a 3D-printed part?
   A) Overly smooth walls
   B) Gaps or missing lines between layers
   C) Excessive stringing
   D) Warped base

29. Why is abrasive flow rate critical in waterjet cutting?
   A) To reduce pump temperature
   B) To increase water pressure
   C) To improve catch tank capacity
   D) For optimal cutting performance and control cost

30. How do fine-feature plasma consumables work?
   A) Increase amperage
   B) Eliminate swirl rings
   C) Reduce shield gas pressure
   D) Reduce nozzle orifice size for tighter arc focus

31. Why is slicer infill overlap important in 3D printing?
   A) To match print color
   B) To improve overhang angles
   C) To increase retraction distance
   D) To ensure perimeter walls bond properly to infill

32. What hazard is unique to laser cutting metals?
   A) Magnetic fields
   B) Reflected beam hazards
   C) Excessive spatter
   D) Hydrogen embrittlement

33. Why would you use a sequenced welding approach on a large assembly?
   A) Increase heat input
   B) Speed up welding
   C) Control distortion and manage residual stress
   D) Reduce filler metal use

34. What's a key reason to apply post-processing to a multi-material assembly?
   A) Improve electrical conductivity
   B) Match surface finishes across materials.
   C) Lower material costs
   D) Reduce part count

35. What does Z-hop in 3D printing help prevent?
   A) Warping on the build plate
   B) Nozzle collisions during travel moves
   C) Cooling fan failure
   D) Print speed errors

36. How can waterjet cutting be adapted for thick composite panels?
   A) Use low-pressure piercing and gradual ramp-up
   B) Eliminate abrasive
   C) Increase abrasive size
   D) Switch to plasma

37. Why use a laser with variable pulse duration?
   A) Optimize material interaction for engraving depth and cut quality
   B) Adjust beam wavelength
   C) Increase machine size
   D) Reduce maintenance cycles

38. What does a welding procedure specification (WPS) provide?
   A) Exact machine settings
   B) Standardized parameters and techniques to ensure quality
   C) Only filler metal types
   D) Consumable brand recommendations

39. How does CAD/CAM integration improve hybrid workflows?
   A) Reduces design complexity
   B) Lowers material cost
   C) Eliminates need for post-processing
   D) Provides a seamless transition between design, toolpathing, and machining

40. What's one reason to use a rotary attachment on a laser system?
   A) Cut thicker plates
   B) Engrave cylindrical objects with precise rotation
   C) Reduce machine cost
   D) Change beam color

41. In CNC routing, why is collet maintenance important?
A) Reduces spindle noise
B) Allows larger tool diameters
C) Increases cutting feedrate
D) Ensures accurate tool holding and reduces runout

42. How to reduce delamination when waterjeting composites?
A) Eliminate abrasive entirely
B) Increase abrasive flow dramatically
C) Pierce slowly and ramp up cutting speed
D) Use higher pressure only

43. What does bridging mean in 3D printing?
A) Printing over unsupported gaps
B) Increasing layer adhesion
C) Changing filament diameter mid-print
D) Reinforcing the build plate

44. How can you reduce plasma arc start failures?
A) Skip consumable replacement
B) Decrease gas pressure
C) Keep torch parts clean and use proper settings
D) Increase arc length

45. Why use multiple passes in laser cutting thick materials?
A) Improve edge quality
B) Increase shield gas cooling
C) Change material hardness
D) Reduce mirror cleaning needs

46. How do you prevent hydrogen-induced cracking?
A) Increase travel speed
B) Preheat and use low-hydrogen consumables
C) Switch to AC current
D) Use no filler metal

47. What's the benefit of combining subtractive (CNC) and additive (3D print) methods?
A) Increase project cost
B) Enable complex shapes and precision features
C) Reduce spindle life
D) Avoid using CAM software

48. How can you improve adhesion on a 3D printer's build plate?

A) Lower nozzle temperature
B) Disable bed leveling
C) Use adhesives, textured surfaces, or heated beds
D) Increase fan speed

49. What's one way laser beam polarization affects cutting?

A) Controls cut width on reflective materials
B) Reduces heat-affected zone
C) Increases machine size
D) Changes laser gas flow

50. What is a critical inspection point after welding aluminum?

A) Ensuring rapid cooling
B) Checking for porosity and verifying oxide removal
C) Confirming part magnetism
D) Checking surface color only

51. Why is G-code simulation important before running hybrid jobs?

A) Visualize toolpaths and detect errors before machine execution
B) Increase spindle power
C) Reduce material thickness
D) Eliminate programming needs

52. How does abrasive hardness impact waterjet performance?

A) Softer abrasives cut faster
B) Harder abrasives improve cutting power and lifespan
C) Abrasive hardness has no effect
D) Harder abrasives lower water pressure

53. What's the risk of excessive retraction in 3D printing?

A) Larger layer height
B) Increased stringing
C) Higher print temperatures
D) Nozzle jamming or under-extrusion

54. Why might you use nitrogen instead of air in plasma cutting stainless steel?

A) Increase oxidation
B) Decrease cutting speed
C) Lower cutting amperage
D) Improve cut edge quality and reduce discoloration

55. What advantage does a fiber laser have for marking metals?

A) Longer wavelength for deep cuts
B) Lower cutting temperature
C) Higher power consumption
D) High-speed, precise, non-contact marking without consumables

56. How can you control distortion in welded assemblies?

A) Weld continuously in one direction only
B) Use balanced welding patterns, clamps, and minimize heat input
C) Reduce shielding gas
D) Increase electrode diameter

57. How does a filament runout sensor help in 3D printing?

A) Adjusts bed leveling
B) Pauses the print when filament runs out to avoid failed prints
C) Speeds up printing
D) Increases infill percentage

58. Why is water filtration important in waterjet systems?

A) Reduce table noise
B) Prevent pump and nozzle damage from contaminants
C) Increase abrasive size
D) Reduce pierce delay

59. What's the main difference between GTAW and GMAW welding?

A) GTAW uses a consumable electrode; GMAW uses a non-consumable
B) GTAW is faster for thick materials
C) GMAW uses a consumable wire feed; GTAW uses a non-consumable tungsten electrode
D) GMAW produces no spatter

60. What does a material library in CAM software provide?
    A) Design templates
    B) Machine diagnostics
    C) Predefined cutting speeds, feeds, and parameters for various materials
    D) File compression

61. Why is regular machine calibration important for CNCs?
    A) Increase spindle RPM
    B) Maintain dimensional accuracy and repeatability
    C) Reduce electricity bills
    D) Avoid using CAM software

62. What's a typical sign of nozzle clogging in a waterjet?
    A) Smoother edge finish
    B) Increased abrasive flow
    C) Decreased cutting power or erratic jet behavior
    D) Shorter pierce time

63. How does layer cooling affect overhangs in 3D printing?
    A) Cooling makes overhangs droop
    B) Cooling reduces print resolution
    C) Cooling has no effect
    D) Proper cooling solidifies layers quickly, improving overhang performance

64. What role does an intensifier play in a waterjet system?
    A) Reduce nozzle diameter
    B) Convert hydraulic power into ultra-high water pressure
    C) Increase abrasive flow
    D) Increase table speed

65. Why is a water table used in CNC plasma cutting?
    A) Minimize fumes, noise, and thermal distortion
    B) Reduce machine weight
    C) Eliminate shielding gas needs
    D) Increase amperage

66. What's a challenge when welding titanium or exotic alloys?
    A) Excessive spatter
    B) High reactivity requiring shielding from air
    C) Too little heat input
    D) Lack of filler materials

67. How does print speed affect 3D print quality?
A) Higher speed usually improves detail
B) Slower speed can improve surface finish and accuracy
C) Speed has no effect
D) Only affects filament color

68. Why is taper compensation needed in waterjet cutting?
A) Adjusts nozzle alignment
B) Increases abrasive efficiency
C) Corrects for angled sidewalls caused by jet lag
D) Reduces water hardness

69. What is a heat-affected zone (HAZ)?
A) The area cooled by water
B) The area near the weld that has undergone microstructural changes due to heat
C) The gas-shielded zone
D) The weld bead only

70. How does proper file naming and version control help hybrid workflows?
A) Prevents errors and ensures everyone uses the correct design files
B) Speeds up machine cutting
C) Reduces spindle load
D) Simplifies cooling system setup

71. What is the advantage of direct-drive extruders in 3D printing?
A) Better handling of flexible filaments
B) Lower weight on the print head
C) Higher maximum print speeds
D) No need for retraction settings

72. What's the main benefit of fume extraction systems in laser cutting?
A) Improve cut quality and protect worker health
B) Increase shield gas flow
C) Change laser wavelength
D) Reduce table weight

73. Why is peening sometimes used after welding?
A) Increase joint stiffness
B) Reduce residual tensile stresses and improve fatigue resistance
C) Decrease weld color
D) Improve magnetic properties

74. How can automated part handling improve multi-machine workflows?
A) Reduce machine costs
B) Minimize manual labor, increase consistency, and reduce cycle times
C) Increase tool wear
D) Eliminate the need for nesting

75. Why is a bowden setup lighter than a direct-drive extruder?
A) The extruder motor is mounted on the frame, not the moving head
B) It uses thinner filaments
C) It reduces print temperature
D) It has no moving parts

76. What's the main risk of overcutting with a laser?
A) Excessively polished edges
B) Burnt edges, excessive kerf, or fire hazards
C) Reduced cutting speed
D) Increased material thickness

77. What are key parameters in hybrid welding (laser + arc)?
A) Only laser power
B) Arc polarity only
C) Balance of laser energy, arc energy, and travel speed
D) Filler metal type only

78. How can modular software improve cross-process integration?
A) Increase machine weight
B) Provide specialized, interoperable tools that communicate across systems
C) Reduce CAD complexity
D) Eliminate toolpath generation

79. What can happen if plasma torch height is too high?
   A) Cleaner cut edges
   B) Wider kerf, poor cut quality, and more dross
   C) Increased cut speed
   D) Decreased HAZ

80. Why is proper storage of welding electrodes important?
   A) Reduce filler metal cost
   B) Prevent moisture pickup, which can cause porosity or hydrogen cracking
   C) Improve color matching
   D) Eliminate shielding gas

81. Why is balanced nesting important in CNC workflows?
   A) Distributes wear evenly across the machine bed
   B) Improves cooling fan speed
   C) Reduces spindle RPM
   D) Allows faster machine upgrades

82. What does “lead-in” mean in a CNC cutting toolpath?
   A) The initial entry into the material to start the cut
   B) The pause before cutting begins
   C) The length of the cutting tool
   D) The material hardness

83. What’s one advantage of SLA (resin) printing over FDM?
   A) Higher print speeds
   B) Fewer post-processing steps
   C) Finer detail and smoother surface finish
   D) No need for support structures

84. What problem does warping create in 3D prints?
   A) Reduced filament flexibility
   B) Increased infill percentage
   C) Over-extrusion at corners
   D) Poor bed adhesion causing edges to lift or curl

85. What role does abrasive recycling play in waterjet cutting shops?
   A) Improve water pressure
   B) Reduce machine noise
   C) Increase abrasive hardness
   D) Reduce abrasive costs and environmental waste

86. Why are proper ground connections critical in welding?
   A) Ensure stable arc and prevent electrical hazards
   B) Reduce shielding gas needs
   C) Increase electrode lifespan
   D) Improve weld color

87. How can CAM software optimize multi-axis CNC machining?
   A) Increase material waste
   B) Generate efficient toolpaths for complex shapes and reduce machine time
   C) Increase operator manual input
   D) Simplify CAD model creation

88. What is key to successful laser engraving on metals?
   A) Apply marking sprays or use fiber lasers for better contrast
   B) Only use CO_2 lasers
   C) Increase material thickness
   D) Cut with maximum laser speed

89. Why should you use a water separator in plasma cutting systems?
   A) Increase cutting amperage
   B) Remove moisture from compressed air to protect consumables and improve cut quality
   C) Change arc polarity
   D) Reduce material hardness

90. What does interpass cleaning achieve in multi-pass welding?
   A) Improves joint alignment
   B) Reduces gas pressure
   C) Removes slag or oxides to ensure strong metallurgical bonds between passes
   D) Increases torch speed

91. What's one reason to use an offset in CNC programming?
   A) Speed up tool changes
   B) Reduce machine weight
   C) Adjust for tool diameter and achieve accurate final dimensions
   D) Increase spindle cooling

92. How does multi-material 3D printing affect design?

A) Makes support structures unnecessary
B) Eliminates need for slicer software
C) Reduces print resolution
D) Requires planning for interface bonding and material compatibility

93. Why is cut taper more noticeable in thicker plasma-cut materials?

A) The arc spreads wider at greater depths
B) The arc becomes more focused at depth
C) Cutting gas cools faster
D) The material becomes magnetic

94. What's the main advantage of using a hybrid CAM system?

A) Only supports plasma systems
B) Integrates toolpath generation across additive and subtractive processes
C) Removes the need for nesting
D) Decreases machine tolerance

95. How does laser duty cycle affect machine lifespan?

A) Only affects beam color
B) Longer duty cycles always improve performance
C) Higher duty cycle reduces lifespan by overheating components
D) Lower duty cycle increases cutting amperage

96. What is a benefit of cold wire feed in GTAW welding?

A) Increase spatter
B) Reduce arc length
C) Improve deposition rates and control
D) Eliminate shielding gas

97. Why is software simulation valuable for multi-process production?

A) Identify collisions, optimize sequencing, and reduce setup time
B) Increase machine power
C) Speed up slicing
D) Reduce design file size

98. What's one challenge when combining 3D-printed and CNC-machined parts?

A) Achieving precise fit due to different dimensional tolerances and surface finishes
B) Reducing filament costs
C) Eliminating the need for adhesives
D) Simplifying CAD modeling

99. Why is surface preparation critical before laser engraving metal?

A) Increases material thickness
B) Ensures clean, consistent marks by removing contaminants
C) Improves laser wavelength
D) Reduces machine noise

100. What's the key factor when estimating timelines for hybrid fabrication projects?

A) Only material availability
B) Only total material cost
C) Only machine weight
D) Accurate breakdown of each process's setup, operation, and post-processing time

You did it!

If this is your first run, congratulations. You just conquered the 100-question multiple-choice quiz. If this is your second run, congratulations again. You clearly enjoy living on the edge to see just how much you've learned from this book.

Now check your answers to see how you did, compare your understanding with the more detailed explanations provided, and record your results at the end of the answer key.

✖ Answer Key and Explanations

Let's see how much you learned from this book!

1B – A compression bit has a unique geometry with opposing cutting flutes, up-cut at the tip and down-cut at the top, pulling material toward the center. This minimizes tear-out on both surfaces, especially in plywood or laminates.

2D – Angular particles like crushed garnet have sharp edges that cut more effectively than rounded ones, leading to faster, cleaner cuts.

3B – Dual extruders can feed different filaments simultaneously, allowing combinations like support and build material or multiple colors in a single print.

4C – Downdraft tables pull smoke, metal dust, and fumes downward and away from the work area, enhancing safety and air quality.

5A – Laser mirrors guide the beam from the source to the lens. Misalignment causes power loss and inconsistent cuts across the bed.

6A – Back-purging with inert gas like argon shields the weld's backside, preventing oxidation that weakens corrosion resistance.

7A – Kerf is the width of material removed by the cutter. Compensation adjusts the toolpath to maintain dimensional accuracy.

8A – Direct drive pumps are simpler and suitable for lighter work, while intensifier pumps generate higher pressures for tougher materials.

9B – Heated chambers maintain consistent temperatures, reducing thermal gradients that cause warping or layer separation, especially with high-temp plastics.

10C – Arc voltage correlates with the torch's height above the workpiece. Automatic sensing ensures optimal distance for clean, accurate cuts.

11C – The CO_2 laser melts the acrylic cleanly, and the molten edge cools quickly to a glass-like, polished finish. Unlike a CNC router.

12A – Porosity occurs when shielding gas is inadequate or contamination is present, causing trapped gas bubbles in the solidifying weld.

13B – Modular jigs standardize part positioning, enabling accurate transitions between different fabrication stages (e.g., cutting, welding, machining).
14D – Runout means the spindle or tool isn't perfectly centered, leading to uneven cuts, tool wear, and poor finish quality.
15D – Coarse garnet (e.g., 60 mesh) cuts faster but may leave rougher edges. Finer mesh improves edge quality but slows cutting.
16A – Brims and rafts create a larger surface area at the base, helping the print stick better to the bed and reduce warping.
17A – Maintaining optimal height ensures the arc is focused correctly, prolongs consumable life, and provides cleaner cuts.
18C – A smaller spot concentrates energy into a finer area, allowing for sharper lines and more intricate engraving details.
19B – Welding aluminum poses unique challenges due to the oxide layer and carries the same high UV exposure risks as other arc welding processes.
20B – Software nesting reduces material costs by packing parts tightly to minimize scrap.
21C – A downcut bit pulls fibers downward, producing cleaner top edges on the material.
22B – A mixing tube combines abrasive with the high-pressure water jet to generate cutting power, allowing the waterjet to cut through hard materials like metal, stone, and composites.
23A – Retraction settings control filament pull-back during non-print moves, which helps reduce stringing and oozing between parts of a print, leading to cleaner results.
24B – Using a worn nozzle or electrode increases the risk of double arcing in plasma cutting because the damaged components disrupt the plasma stream and cause arc instability.
25C – A fiber laser is more efficient at cutting reflective metals like aluminum, copper, and brass compared to a CO_2 laser, while also offering higher electrical efficiency and lower maintenance needs.

26B – Excessive convexity or underfill in a weld bead can weaken the joint by creating stress concentrations and reducing proper fusion, directly affecting mechanical strength.

27C – Using precise fixturing and maintaining tolerances ensures cross-machine precision when combining CNC and welding, allowing accurate alignment and consistent results.

28B – Gaps or missing lines between layers are a clear sign of under-extrusion in a 3D-printed part, indicating insufficient filament flow.

29D – Abrasive flow rate is critical in waterjet cutting because it ensures optimal cutting performance while controlling abrasive consumption and overall operating costs.

30D – Fine-feature plasma consumables work by reducing the nozzle orifice size, which tightens arc focus and improves cut precision on detailed features.

31D – Slicer infill overlap ensures that perimeter walls bond properly to the infill, improving part strength and preventing internal separation or weak spots.

32B – Reflected beam hazards are unique to laser cutting metals, since shiny or reflective surfaces like aluminum and copper can redirect the laser beam, posing serious safety risks.

33C – A sequenced welding approach is used on large assemblies to control distortion and manage residual stress by balancing heat input across the structure.

34B – Post-processing a multi-material assembly helps match surface finishes across materials, creating a consistent appearance and improving product quality.

35B – Z-hop in 3D printing helps prevent nozzle collisions during travel moves by lifting the nozzle slightly when moving over already printed areas.

36A – Waterjet cutting thick composite panels can be adapted by using low-pressure piercing and gradual ramp-up to reduce delamination and ensure a clean cut.

37A – A laser with variable pulse duration allows optimization of material interaction for engraving depth and cut quality, improving precision across different materials.

38B – A welding procedure specification (WPS) provides standardized parameters and techniques to ensure consistent weld quality and compliance with industry standards.

39D – CAD/CAM integration improves hybrid workflows by providing a seamless transition between design, toolpathing, and machining, reducing errors and setup time.

40B – A rotary attachment on a laser system allows engraving cylindrical objects with precise rotation, ensuring accurate and consistent markings around the circumference.

41D – Collet maintenance is important in CNC routing because it ensures accurate tool holding and reduces runout, which improves cut quality and tool life.

42C – To reduce delamination when waterjeting composites, pierce slowly and ramp up cutting speed to minimize stress on the layers and maintain a clean cut.

43A – Bridging in 3D printing refers to printing over unsupported gaps, allowing the filament to span gaps without sagging or collapsing.

44C – Plasma arc start failures can be reduced by keeping torch parts clean and using proper settings, ensuring reliable arc initiation and consistent cutting performance.

45A – Multiple passes in laser cutting thick materials improve edge quality by gradually removing material and reducing heat-affected zones.

46B – Hydrogen-induced cracking can be prevented by preheating the workpiece and using low-hydrogen consumables to minimize hydrogen absorption in the weld.

47B – Combining subtractive (CNC) and additive (3D printing) methods enables creation of complex shapes with precise features that would be difficult or impossible with either method alone.

48C – Adhesion on a 3D printer's build plate can be improved by using adhesives, textured surfaces, or heated beds to help the first layer stick properly.

49A – Laser beam polarization affects cutting by controlling cut width on reflective materials, improving efficiency and precision when cutting metals like aluminum or copper.

50B – After welding aluminum, it is critical to check for porosity and verify oxide removal to ensure weld integrity and prevent weak spots.

51A – G-code simulation is important before running hybrid jobs to visualize toolpaths and detect errors before machine execution, reducing the risk of collisions or mistakes.

52B – Harder abrasives improve waterjet cutting power and extend abrasive lifespan, enabling cleaner cuts through tough materials.

53D – Excessive retraction in 3D printing can cause nozzle jamming or under-extrusion, disrupting filament flow and print quality.

54D – Nitrogen is used instead of air in plasma cutting stainless steel to improve cut edge quality and reduce discoloration caused by oxidation.

55D – A fiber laser offers high-speed, precise, non-contact marking on metals without consumables, making it efficient and low-maintenance.

56B – Distortion in welded assemblies can be controlled by using balanced welding patterns, clamps, and minimizing heat input to manage residual stress.

57B – A filament runout sensor helps by pausing the print when filament runs out, preventing failed prints and wasted material.

58B – Water filtration in waterjet systems is important to prevent pump and nozzle damage from contaminants, ensuring consistent performance and longer component life.

59C – The main difference is that GMAW uses a consumable wire feed, while GTAW uses a (technically) non-consumable tungsten electrode, affecting process control, speed, and applications.

60C – A material library in CAM software provides predefined cutting speeds, feeds, and parameters for various materials, helping optimize machining performance and reduce trial-and-error setup.

61B – Regular machine calibration is important for CNCs to maintain dimensional accuracy and repeatability, ensuring consistent, high-quality parts.

62C – A typical sign of nozzle clogging in a waterjet is decreased cutting power or erratic jet behavior, which can affect cut quality and consistency.

63D – Proper layer cooling solidifies layers quickly in 3D printing, improving overhang performance and reducing sagging or drooping.

64B – An intensifier in a waterjet system converts hydraulic power into ultra-high water pressure, enabling the jet to cut through tough materials.

65A – A water table is used in CNC plasma cutting to minimize fumes, noise, and thermal distortion, improving safety and cut quality.

66B – Welding titanium or exotic alloys is challenging because of their high reactivity, requiring shielding from air to prevent contamination and embrittlement.

67B – Slower print speed in 3D printing can improve surface finish and dimensional accuracy by allowing better layer adhesion and precise filament deposition.

68C – Taper compensation in waterjet cutting is needed to correct for angled sidewalls caused by jet lag, ensuring parts meet dimensional specifications.

69B – A heat-affected zone (HAZ) is the area near the weld that has undergone microstructural changes due to heat, which can affect material properties and strength.

70A – Proper file naming and version control in hybrid workflows prevents errors and ensures everyone uses the correct design files, maintaining consistency and reducing costly mistakes.

71A – Direct-drive extruders offer better handling of flexible filaments by reducing the distance between the drive gear and the hotend, improving control and reducing filament buckling.

72A – Fume extraction systems in laser cutting improve cut quality and protect worker health by removing smoke, particulates, and harmful gases from the work area.

73B – Peening is sometimes used after welding to reduce residual tensile stresses and improve fatigue resistance, helping prevent cracks in the weld area.

74B – Automated part handling improves multimachine workflows by minimizing manual labor, increasing consistency, and reducing cycle times for higher efficiency.

75A – A Bowden setup is lighter than a direct-drive extruder because the extruder motor is mounted on the frame instead of the moving print head, reducing moving mass.

76B – *The main risk of overcutting with a laser is burnt edges, excessive kerf, or fire hazards, which can damage the part and surrounding material.*

77C – *Key parameters in hybrid welding (laser + arc) include balancing laser energy, arc energy, and travel speed to achieve optimal penetration, bead shape, and weld quality.*

78B – *Modular software improves cross-process integration by providing specialized, interoperable tools that communicate across systems, streamlining hybrid workflows and reducing errors.*

79B – *If plasma torch height is too high, it can cause a wider kerf, poor cut quality, and more dross, reducing precision and edge finish.*

80B – *Proper storage of welding electrodes prevents moisture pickup, which can cause porosity or hydrogen-induced cracking in welds.*

81A – *Balanced nesting in CNC workflows distributes wear evenly across the machine bed, improving part quality and extending tool and machine life.*

82A – *A "lead-in" in a CNC cutting toolpath is the initial entry into the material to start the cut smoothly and prevent defects at the cut start.*

83C – *SLA (resin) printing offers finer detail and smoother surface finish compared to FDM, making it ideal for intricate models and prototypes.*

84D – *Warping in 3D prints causes poor bed adhesion, leading to edges lifting or curling, which can distort the part and ruin the print.*

85D – *Abrasive recycling in waterjet cutting shops reduces abrasive costs and environmental waste by reusing material safely and efficiently.*

86A – *Proper ground connections in welding ensure a stable arc and prevent electrical hazards, protecting both the operator and the equipment.*

87B – *CAM software optimizes multi-axis CNC machining by generating efficient toolpaths for complex shapes, reducing machine time and improving precision.*

88A – *Successful laser engraving on metals relies on applying marking sprays or using fiber lasers to achieve better contrast and durable markings.*

89B – *A water separator removes moisture from compressed air, protecting consumables and improving cut quality.*

90C – Interpass cleaning in multi-pass welding removes slag or oxides between passes, ensuring strong metallurgical bonds and consistent weld quality.
91C – Using an offset in CNC programming adjusts for tool diameter, helping achieve accurate final dimensions and precise part geometry.
92D – Multi-material 3D printing affects design by requiring planning for interface bonding and material compatibility to ensure structural integrity and proper adhesion between materials.
93A – Cut taper is more noticeable in thicker plasma-cut materials because the arc spreads wider at greater depths, creating angled sidewalls.
94B – The main advantage of using a hybrid CAM system is that it integrates toolpath generation across additive and subtractive processes, streamlining workflows and reducing errors.
95C – Laser duty cycle affects machine lifespan because a higher duty cycle can overheat components, reducing the overall longevity of the system.
96C – Cold wire feed in GTAW welding improves deposition rates and control by allowing precise addition of filler metal without affecting arc stability.
97A – Software simulation is valuable for multi-process production because it identifies collisions, optimizes sequencing, and reduces setup time, improving efficiency and safety.
98A – One challenge when combining 3D-printed and CNC-machined parts is achieving precise fit due to different dimensional tolerances and surface finishes between the processes.
99B – Surface preparation before laser engraving ensures clean, consistent marks by removing contaminants like oils or oxidation.
100D –Estimating timelines for hybrid fabrication requires breaking down each process's setup, operation, and post-processing time.

Quiz Score BEFORE reading this book: ***____/100***

Score AFTER reading this book: ***____/100***

Final Words
The Build Never Really Ends
Stay Curious and Iterate Boldly

If you've made it all the way to the end of this book as well, first, congratulations yet again. You've waded through pages of laser beams, plasma arcs, high-pressure waterjets, router bits, extruded filament, and molten metal.

You've tackled the why, the how, and the "oh no, I forgot to clamp that" moments that make up life in a modern fabrication shop. And guess what? You're still standing. That alone earns you major shop cred.

So what are the final words I want to leave you with? Simple: the build never really ends.

That's not just a catchy phrase, it's a hard truth in this world. Whether you run a giant industrial shop, a one-person operation like myself at *CNCROi.com*, or just tinker when you can, there's always another challenge waiting.

Machines evolve, materials shift, clients bring weirder requests, and your own ambitions start whispering, "Hey, what if we tried something totally different?" One day, you're cutting mild steel; the next, you're figuring out how to 3D print using wood filament or precision-engrave titanium. That's the joy, and the headache, of this craft.

Importantly, the real tool that keeps you sharp isn't your CNC router, your shiny new laser, or that plasma cutter you just convinced yourself was a "necessary investment." It's your mindset. Specifically: your curiosity.

The best fabricators, machinists, and makers I've ever met all share one thing: they never stop asking questions. There is always one more follow-up question needing answered.

They ask why something works, why it does not, and how to make it better. I guess entrepreneurship is the same way, a curiosity that can never be fully satisfied.

They never assume they know everything, never stop poking at the limits, and never stop hunting for better, smarter, or just plain weirder ways to solve a problem.

And yes, they also know when to laugh at their own mistakes, because if you're not laughing at least some of the time, you're probably taking this all too seriously.

This Q&A book was written not as a definitive rulebook, but as a companion. A friendly guide on your bench to remind you that it's okay not to know everything, as long as you're willing to dig in, test, fail, and try again.

So, here's your final nudge: stay curious. Stay flexible. Keep learning. And most importantly, keep making things that bring you (and maybe even your clients) a "spark of joy."

At the end of the day, whether you're cutting, welding, printing, routing, or just sweeping the shop floor, you're part of a tradition that stretches all the way back to the first human who decided, "Hmm, maybe I can shape this rock into something useful."

The tools have changed, the stakes have changed, but the drive? That's eternal. Keep feeding it so it grows with your skills and dreams into something you can look back on fondly, still cracking a drooling grin in your wheelchair at the old folks' home.

Although another CNC book is not on the near horizon, CNC milling is definitely calling my name as are quite a few shop improvements. Since joining the trade, I really want to make a welding projects book though, so much knowledge in that field to share. Keep living through your passions.

There always seems to be one more toy to slip into the toolbox. I know a few old timers who never shook this addiction we call learning, and honestly, I hope it never ends for me either.

Now go on, pester the experts with endless questions like a toddler who just learned to talk, and make something amazing. The build never truly ends, and honestly, would you even want it to?

Appendices
Deeper Dives and Handy Extras
Charts, Specs, and Tools to Keep You Smart

Appendix A: Tool upgrade or investment advice

The seductive allure of the tool upgrade, that moment when your shop-floor heart starts beating faster as you eye a new CNC router, a fancier laser, a bigger plasma cutter, or maybe a 3D printer that claims it can print titanium and cook your breakfast.

But when is the right time to invest? We're not talking about myths or mistakes here, we're talking about real, valid, totally-justifiable, non-impulse reasons to consider pulling the trigger on that shiny, expensive upgrade.

Here's a rundown of 10 honest reasons you should consider investing in a tool upgrade, because sometimes "want" and "need" do actually line up (and sometimes, you can at least pretend they do).

1- Your Current Machine Is Crying for Help

Let's face it: if your machine sounds like a dying robot every time you press start, or if you're performing ritualistic "please work" dances just to coax it through one more job, it's probably time to upgrade.

Machines age, parts wear out, and technology moves on. If you're spending more time fixing your equipment than using it, it's not heroic, it's just inefficient. Save yourself the sweat and the emotional trauma: sometimes, a shiny new machine is the merciful choice.

2- You're Turning Down Work You Could Be Doing

A client asks, "Can you cut this thicker metal?" or "Can you handle this exotic material?" and you sigh, shake your head, and hand the job off to someone else. It's painfull!

If you're consistently turning away jobs you want to take because your equipment can't handle them, it's time to think upgrade. Letting opportunity walk out the door because you're clinging to old limits is basically leaving money on the table, and what shop owner doesn't love money?

3- You Need More Speed

If you're stuck babysitting long runtimes, holding your eyelids open with one hand while the other refills your coffee for the third time, consider whether faster machines might actually help. For instance, getting router with an ATC cut down production of my custom wood banks by 75%!

Modern routers, lasers, plasma tables, and even 3D printers are often dramatically faster (and more efficient) than models from even five years ago. Faster machines mean more throughput, and more throughput means you can do more work, bill more clients, and maybe even go home before midnight once in a while.

Waterjet too slow? Get a fiber laser!

4- You're Expanding Into New Materials or Techniques

Maybe you've been cutting wood for years, but now clients want you to work with aluminum, or acrylic. Or maybe you've mastered 2D cuts but want to move into 3D engraving, hybrid fabrication, or metal additive manufacturing.

Each material or process often requires a tool upgrade, not just because of capability, but because pushing an old machine to do something it wasn't designed for usually ends in smoke.

5- Your Clients Expect Higher Precision or Finish

If you're hearing phrases like "Can you make the edge a little cleaner?" or "We need tighter tolerances for this aerospace project," take the hint. Consumer-grade or entry-level machines often max out on precision and surface quality, but industrial or pro-level tools can hit those fine details effortlessly. Upgrading when your client base evolves ensures you stay competitive, and avoids hours of hand-finishing that eat into your profits.

6- You're Spending Too Much on Outsourcing

Sure, outsourcing tricky jobs can save money, but if you're consistently sending out the same kind of work, you're paying someone else's machine bills (and profit margins) instead of investing in your own shop.

Run the numbers: if you're dropping thousands a year outsourcing waterjet, laser, or large-format 3D printing work, you might actually save money long-term by bringing that capacity in-house. Bonus: you get to control turnaround times, quality, and pricing.

7- You're Moving Into Production, Not Just Prototyping

A trusty entry-level 3D printer or CNC router is great for occasional jobs and prototypes, but when you shift into production, churning out batch after batch, you need something that's built for heavy-duty work.

Projects getting too big? Get a bigger machine!

Production-level tools have stronger frames, faster speeds, better cooling, and components rated for extended runtime. Without them, you're bottlenecking your growth.

If you've made the leap from "fun side project" to "actual business," it's time your tools reflected that.

8 - Software and Automation Have Leveled Up

Modern tools often come bundled with dramatically improved software: better CAM, smarter toolpathing, more automation features, or built-in sensors that make setup faster and errors rarer.

If you're still manually jogging your machine, tweaking offsets by hand, or wrestling with outdated control software, you're wasting time.

An upgrade can reduce setup time, streamline operations, and give you more brainpower to focus on creative or business challenges instead of fighting clunky systems.

9 - You're Running Out of Capacity (and Shop Space)

Here's a weird one: sometimes upgrading reduces your space problems.

If you've got three small machines running half the day just to meet demand, a single larger, faster, more versatile machine might replace all three : freeing up space, cutting maintenance complexity, and improving efficiency.

Plus, you get the satisfaction of consolidating your workflow, which can open up room for (surprise!) more tools down the road.

10 - You Want to Future-Proof Your Shop

Sometimes, you upgrade not because you have to, but because you're thinking ahead. Machines have lifespans, and you don't want to hit a catastrophic failure right when your biggest client needs a rush order.

Planning tool upgrades ahead of breakdowns ensures you have time to research, budget, and integrate new equipment smoothly. This also means you get top dollar for your equipment as value drops once a breakdown happens.

Final Thoughts: Upgrade With Purpose, Not Just Passion

Look, we all get the tool upgrade itch. It's part of the maker DNA, the endless hunger to level up, improve, and tackle bigger, bolder projects. But the smartest upgrades come from matching your investments to real, tangible needs: improving capacity, precision, speed, or capability in ways that unlock new opportunities or solve persistent problems.

So, next time you're drooling over that next-gen plasma table or multi-material 3D printer, don't just ask, "Do I want this?" Ask, "Does this move my shop forward?" If the answer is yes, congratulations: you've earned your upgrade.

Just remember to leave room in the budget for snacks, you're going to need fuel while you watch that beautiful new machine do its magic.

CNC plasmas are a must-have for any welding shop.

Don't let a 5 x 10 ft CNC router table go to waste.

Appendix B: Pricing Digital Fabrication: Why Time Isn't the Only Metric

If you've ever tried to price out a CNC job, whether it's a single plasma-cut bracket or a hundred engraved panels, you've probably defaulted to the easiest metric you know: time. After all, time is money, right?

Just track the machine hours, multiply by an hourly rate, and boom: an invoice is born. But if you've been in this game for more than about five minutes, you already know this method can be wildly misleading.

Pricing based solely on time is like judging a race by how cool the car looks: you'll miss most of what actually matters.

Welcome to the messy, occasionally mind-bending world of pricing digital fabrication. This essay is your practical roadmap through the maze.

We'll talk machine amortization, tool wear, complexity pricing, and the big one: value-based pricing. Because spoiler alert, what something is worth to a customer has very little to do with how long it took you to make it. Ready? Let's talk money.

Hourly Rates: A Reasonable Starting Point (and a Terrible Ending Point)

Look, hourly rates aren't all bad. They're familiar. They give you a sense of structure. And they help ensure you're not working for free, which, I'll be honest, is a problem more makers face than they care to admit.

For freelancers and small shop owners, setting an hourly machine rate is often the first step.

But here's the thing: your hourly rate needs to reflect more than just machine cycle time. It should account for everything that keeps your shop running. Short of being Einstein, there's no way to generate an exact figure from these details.

That includes your overhead: rent, insurance, utilities, and yes, your vacation habit. It includes tool wear, consumables, operator time (even if you're the operator), setup and teardown, waste, mistakes, and the occasional "oh no, the bit snapped" moment.

If you're charging $60/hour because that's what someone said on a forum, but your actual costs are closer to $85/hour, guess what? You're paying customers to let you work.

That's not a business model, it's a slow nose bleed.

Amortization: Paying Off That Shiny Machine

Remember the day you brought your CNC router or laser home? It was glorious. You wiped it down, updated the firmware, and told your significant other it would "pay for itself in no time."

Now's the time to make sure that's actually happening.

Machine amortization is the process of spreading the cost of your equipment over its useful life.

For example, let's say your machine cost $20,000. You expect it to last five years, and you estimate it will run for 1,200 hours per year. That gives you 6,000 total production hours.

So, just to recover the cost of the machine, you need to charge $3.33 per hour. Add in maintenance, software licenses, and upgrades, and that number easily climbs to $5–6 per hour.

And that's before the spindle even spins, the plasma strikes an arc, or the laser decides to self-destruct because you left acrylic cutting in it overnight.

Now multiply that across multiple machines. Your hourly rate isn't just about time, it's about covering the cost of the gear that gives you the time in the first place. These are all things to take into consideration when giving a quote and your numbers, guaranteed, will be wildly different than other shops even if they have the same equipment as you.

Tool Wear and Consumables: Hidden but Deadly

Let's talk bits. And end mills. And nozzles, belts, laser lenses, plasma tips, you know, the stuff that gets chewed up doing the real work.

Tool wear is one of the most overlooked pricing factors in digital fabrication. If you're not factoring in how often your $30 compression bit burns out cutting plywood, you're quietly losing profit with every pass.

Take a $30 bit that lasts for 10 jobs, that's $3 per job. Over 100 jobs, that's $300 straight out of your margins.

Now apply that logic to every consumable: adhesives, protective films, mist coolant, air filters, replacement belts, even the occasional therapy chocolate after a failed print.

Always include a tool wear line in your internal pricing, even if it's small. It adds up. And yes, customers should help pay for your bits. After all, you're wearing them out on their jobs that for some reason, are always due yesterday.

Complexity: Not All Jobs Are Created Equal

Here's a hard truth: not all one-hour jobs are worth the same.

Take Job A: cutting 12-inch round discs out of MDF. It's straightforward: one bit, one sheet, no hold-down issues thanks to that vacuum table you bought on my advice from my previous book.

Now compare that to Job B: machining a complex aluminum bracket with five tool changes, a 3D surface contour, tight tolerances, and two post-process steps.

Both jobs may take an hour of machine time, but should they be priced the same? Of course not.

Because you're not just selling machine time, you're selling complexity, precision, risk, and expertise. The tougher the job, the less competition, thus, your margins should go up.

This is where complexity pricing comes in, a mix of science and intuition. Certain factors should raise your price: multi-tool jobs, tight tolerances, expensive or exotic materials, intricate geometries or long toolpaths, custom fixturing, and any hand-finishing or post-processing.

You can build a complexity multiplier into your pricing model. For example, use a 1.0x rate for simple jobs, 1.25x for moderate complexity, and 1.5x or higher for demanding work. It's not exact math, but it better reflects the real effort involved.

Time alone doesn't tell the full story. Complexity does.

Setup Time and Operator Brainpower

CNC machines may be robotic, but setup is still very human, and often the human part is harder than the cutting itself. Programming, thinking ahead, troubleshooting, and double-checking are not things you can simply automate away. They require skill, experience, and judgment, and they make or break the job before the spindle ever touches material.

If a job takes you 45 minutes to dial in tool changes, Z-height probing, vacuum zones, fixturing, software checks, and G-code simulations, and only 15 minutes to actually run, then pricing by cycle time alone misses 75% of the real labor involved.

That estimate does not even include the small but constant overhead such as walking back and forth to grab a gauge block, re-reading a material spec, or stopping to reclean a surface because dust ruined your vacuum hold-down.

Some shops bill setup time separately. Others amortize it across a batch or apply a flat rate for smaller jobs. The important part is recognizing that setup is not free labor. In fact, it is often where your expertise matters most.

A well-thought-out setup prevents broken tools, scrapped parts, or an expensive crash, which can save hours compared to rushing through the process.

Value-Based Pricing: The Holy Grail (and the Trapdoor)

Here's where things get spicy.

Value-based pricing means charging based on the value you deliver to the customer, not just your cost plus a markup. A replacement part that gets a production line moving again? That's high value. A custom wedding sign for a bride on a tight deadline? High value. A one-off prototype for a $10M medical device? Extremely high value.

This approach lets you price well above your hourly rate, sometimes dramatically so. And yes, it feels strange at first. But this is exactly how high-margin industries operate. You're not just selling time or materials, you're selling a result. And results are all that really matters.

The trap? You need to deeply understand your customer: their deadlines, stakes, and alternatives. It takes empathy, research, and sometimes just asking, "What happens if this part doesn't arrive on time?"

If the answer is "I lose a $50,000 contract," then maybe you shouldn't be charging $50 for that rush job. Just a thought.

When to Say No (or Price to Say No)

Not every job is worth taking, and sometimes, pricing should be a polite way to say no. At *CNCROi.com*, I honestly spend more time saying NO to jobs than YES.

If a client wants one unit, made from a difficult material, using a terrible file, and delivered by yesterday, you have two choices: say no outright, or price it so high it pays for your pain. What is the price for turning your customer's problems into yours?

In industry circles, this is often called "F**k-off pricing." It is no secret; even customers recognize it, since every shop resorts to it now and then. Yes or no, end of discussion.

Either option is valid. Just don't say yes to jobs that burn your time and waste your equipment.

Bundling and Batch Pricing

Batch jobs change the equation. Once the setup and design work are done, the marginal cost of additional units drops significantly, giving you room to discount while still boosting overall profit.

Say setup takes 1 hour, machining time per unit is 10 minutes, and tooling is a flat cost. For one unit, the price needs to be high to cover all that overhead. But for ten units, the per-unit cost drops, and your total profit rises.

Smart shops use this to upsell: "You know, for just $X more, I can make five of these instead of two." Many clients take the deal, and you earn more by doing the same job just a little longer.

Remember, just because a customer did not ask for something or even think of it, that does not mean you should avoid suggesting alternatives that benefit both sides. In many cases, the most valuable upsell you can offer is not a product or service, but the insight to propose smarter solutions.

The right idea can boost your profit margin while also making you a more trusted supplier to a customer who often has needs they have not yet recognized themselves.

Pricing Transparency vs. Pricing Confidence

Do you show your math to clients? That depends.

Some customers appreciate line-item pricing. $85/hour for machine time, $25 for setup, $10 for material, $5 for tool wear. Others just want a clean bottom-line number. Remember, transparency isn't the same as trust. Clients don't need to see every detail, they just need to feel that your pricing is fair, consistent, and professional.

What matters more than how you present your pricing is how you stand behind it. If you're unsure, apologetic, or worse, constantly undercutting yourself, people will pick up on that. Be confident in your pricing.

Tools to Help You Price Smarter

Here are a few ways to systematize your pricing.

Start with spreadsheets, build one that includes your hourly rate, material costs, tool wear, and complexity multipliers.

Use job estimators in tools like Fusion 360 or SheetCAM to get accurate runtime estimates that inform your pricing.

Consider CRM or quote software like Jobber, Core, or even a custom Notion database to track past quotes, pricing, and margins. And don't forget benchmarking: check what similar services charge, not to copy them, but to stay competitive. You might get away with one or two bad quotes, but any more than that and you risk being blacklisted.

Final Thoughts: Know Your Worth, Charge Accordingly

Time matters, yes, but it is just one ingredient in a much richer pricing stew. Good pricing reflects your costs, your tools, your skill, your customer's urgency, and your willingness to take the job. It also reflects your ability to anticipate headaches before they happen, because every hidden complication has a way of turning "cheap" into "expensive" fast when it comes to custom fabrication.

In the end, pricing digital fabrication is not just about plugging numbers into a calculator and hoping for the best. It is about understanding the value you bring, the complexity of the job, and what the client truly needs. It is equal parts math, judgment, and experience, and the balance shifts with every project.

And if all else fails, remember this: never quote a job before going through the mental gymnastics of running it through your head. That is how you end up cutting 1" thick aluminum at 1/8" MDF prices.

Charge smart. Build better. And may your bits stay sharp, your margins stay healthy, and your invoices paid on time!

Guitar templates CNC laser cut.

Guitar templates CNC router cut.

Appendix C: CNC as a Side Hustle: Pitfalls and Possibilities

So you've got a CNC router in your garage, a 3D printer on the kitchen table, and a dream of turning weekend cutting sessions into ongoing cash. Welcome to the glorious world of side hustling in custom digital fabrication.

Maybe you want to make some extra money. Maybe you're testing the waters before quitting your day job. Maybe you just want your tools to pay for their own upgrades (and maybe a few pizza-fueled late nights). Whatever your reason, starting a CNC side hustle can be wildly rewarding.

Let's break down the real-world realities of running CNC as a side gig: the pros, the traps, the "wish I'd known that" moments, and how to go from hobbyist to semi-pro without losing your sanity (or your weekends).

Why a Side Hustle Makes Sense

Let's start with the upside: CNC machines are built for repeatability. If you can make one good item, you can usually make ten, or a hundred. Unlike many hobbies, digital fabrication scales surprisingly well, making it a natural fit for part-time income.

The benefits are solid. There's a low barrier to entry, especially if you already own the machine. Custom, precision-cut parts have high perceived value. You can work flexible hours around your day job. You get creative freedom to build what excites you. And there's real potential to scale into a full-time business down the road.

So far, so good. But as you'll soon see, the side hustle path isn't all cash and chips.

Trap #1: Undervaluing Your Time

It's easy to price like a hobbyist when you're just getting started. You think, "I'm doing this on weekends anyway, and it only took two hours, so maybe $50?" Wrong.

You have to price like a business, even if you're not one yet. Why? Because you're still using electricity, bits, belts, and software licenses. You're still burning time you could spend elsewhere. And you're setting customer expectations that will haunt you later.

Start by calculating your real costs and set a minimum viable rate. Don't sell below it, even if you're just "doing it for fun." You are a business, not a charity.

Trap #2: Taking Any Job That Comes Along

The thrill of landing your first paid job is undeniable. That first order feels like validation, proof that someone values your work enough to pay for it. But once the excitement wears off, it is easy to fall into the trap of saying yes to absolutely everything. Signs for a dog groomer, bulk wedding keychains, or even a one-off bracket for someone's 1987 motorcycle that they insist must be machined from solid gold. Automotive jobs are ALWAYS trouble BTW.

When that happens, you are no longer building a business. You are building a time sink that drains your energy, and leaves you feeling like an unpaid intern to your own shop. Every oddball project pulls you further away from the kind of work that actually builds long-term value.

The antidote is learning to say no, or better yet, learning to specialize early. Choosing a focus area does not limit you, it defines you. Whether it is home décor, small parts prototyping, guitar inlays, architectural models, or custom panels for cars and motorcycles, specialization helps you refine your skills and build a recognizable identity.

When you are known for something specific, you stop competing for scraps and start attracting jobs that fit your shop and your expertise. Better jobs mean better margins, and better margins usually come with better customers—the kind who respect your time and trust your recommendations.

In short, you stop being "someone with a machine" and start being the local go-to person in your field.

Trap #3: Thinking the Machine Will Do the Work

This one stings. People assume the CNC will run itself while they do other things. But if you've actually used one, you know better. Or like Simon, who burned down his shop.

There's prep: CAD, CAM, material sizing, bit selection. Then setup: tool changes, probing, clamping. After that, cleanup: dust, tabs, part finishing. And always, troubleshooting, like figuring out why the spindle just drove through the table. Again. For the 4th time, after getting through 95% of the job.

The machine is a multiplier, not a magic box. You still need to plan for it, supervise it, and clean up after it.

Trap #4: Underestimating Design Time

Design is invisible labor. Customers don't see you iterating in Fusion or fixing a messy .dxf from someone's cousin. But you feel it.

Design time is real time. Bill for it. And if a client sends you "a quick sketch" on a napkin and wants you to "just make it work," you're doing engineering, not just fabrication.

Tip: Separate design fees from cutting fees in your quotes. It sets boundaries and makes you look more professional.

Trap #5: Growing Without Systems

At first, you remember every setting and file location. Then version 3, 4, and 5 of the same sign roll in. Six months later, a customer wants a reorder, what font did you use? Where's the G-code? Which bit? Without systems, growth turns into chaos.

Start documenting from day one. Keep job sheets with setup notes. Use clear file naming conventions. Maintain tool libraries with notes. Log your maintenance. Build quote templates. These don't need to be fancy, a simple Google Doc or Notion page will outperform your memory every time. I generally delete the files, and put this burden on the client to make sure I'm working from the latest version.

The Possibilities: Turning Skill Into Sales

Okay, enough doom and gloom, here's the good stuff: people love well-made, custom parts. And you have a magical machine that can make almost anything. That gives you power, and serious profit potential.

What can you sell? You can offer products like signage, decor, nameplates, furniture parts, jigs, inserts, enclosures, or cosplay gear.

You can provide services such as cutting, engraving, prototyping, finishing, part replication, or reverse engineering.

You can create kits like flat-pack furniture, laser-cut models, or modular organizers. You can even sell design files on platforms like Etsy or Gumroad.

Or go hybrid: combine laser cutting, 3D printing, CNC, or welding into one-of-a-kind offerings.

Find what you enjoy making and what actually sells, then refine your process until it's repeatable.

Workflow Tips for Part-Time Shops

Batch everything by grouping similar jobs together to minimize tool changes and setup time.

Schedule rigid time blocks and treat your shop time like an appointment, don't let distractions creep in.

Build quote templates so you can reuse common pricing structures quickly and consistently.

Track every job with a simple spreadsheet that records what you made, how long it took, and what you earned.

And automate where you can, using CAM templates, post-processors, and pre-made shipping labels to save clicks and time.

When to Scale (and When Not To)

Some side hustles stay part-time forever, and that's perfectly fine. But if you're thinking about going full-time, make sure you're ready.

Scale when you're turning down work, when you've built a repeatable workflow, and when you have a product that's consistently profitable.

Don't scale if you hate customer service, if you're buried in unprofitable jobs, or if you still haven't nailed your niche.

Going full-time means taking on full-time responsibility. It's not just more making, it's a business, with taxes, insurance, deadlines, and customers who expect you to deliver. Again, this is night and day compared to being an employee. Not everybody can handle this stress.

Legal, Financial, and Boring (But Necessary) Stuff

Register a business name, even if it's just a sole proprietorship. Get liability insurance, especially if people will be using your parts. Keep your finances clean by separating bank accounts for income and expenses, and track every cost: bits, materials, software, tools.

File your taxes properly, you might owe income tax, but you could also qualify for valuable write-offs.

None of this is glamorous, but it protects you. And when your business is legit, people take you more seriously.

Set realistic goals. Don't expect to make $10K a month in your first three months. Instead, aim to cover your machine costs within the first year.

Don't burn out by working every weekend and night. Focus on projects that are interesting and rewarding.

Remember: this is a side hustle. It should support your life, not consume it. I know, easier said than done, especially if you are passionate about what you are doing, as I am.

Final Thoughts: Respect the Hustle, Master the Craft

CNC side hustling isn't a get-rich-quick plan. It's a craft-based business with some serious growth potential but that's what makes it great. You're not dropshipping junk. You're building real things with real value.

Yes, it can get messy. Yes, you'll underquote, overcut, and forget to charge for shipping at some point. But you'll learn. You'll refine. You'll get better. And if you play it smart, price it fairly, and focus your energy, you'll find that your shop can do more than just make cool parts.

It can make income. It can make connections. It can make your weekends a little more productive and a lot more fun.

But remember, the path to business success is a roller-coaster full of unpredictable challenges. Pandemics, financial crises, and whatever else might be lurking around the corner can affect you both directly and indirectly. It is far from all rainbows and sunshine.

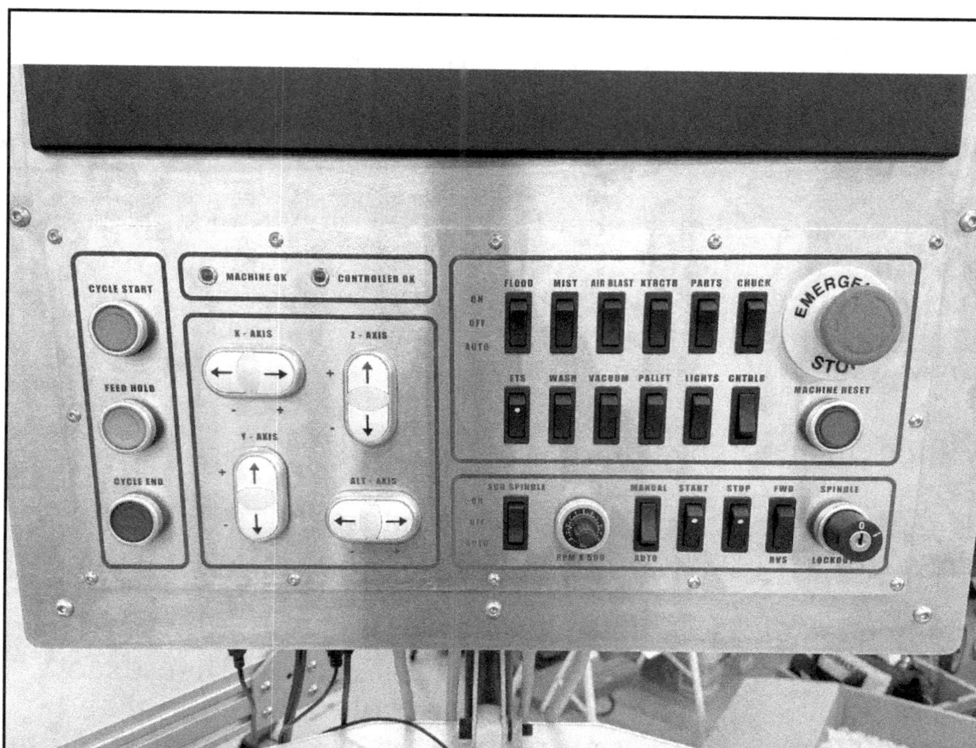

Fiber laser cut and etched stainless steel control panel.

Appendix D: Standing Out in a World of Sameness: Finding Your CNC Niche

There are millions of CNC platforms out there, from routers and lasers to 3D printers, plasma cutters, and just about anything else you can imagine. They're constantly at work: cutting parts, engraving logos, printing widgets, and fueling an endless stream of Etsy listings. In fact, even your everyday desktop printer technically counts as a CNC.

So here's the big question:

Why should anyone buy from you?

That's not meant to discourage you. It's meant to sharpen your thinking. Because in a world where anyone with $3,000 and a garage can start a CNC side hustle, standing out isn't just a nice bonus. It's the difference between "Oh yeah, they do signs too" and "They're the ones who make those beautiful hardwood topographic maps."

Let's deep dive into finding your CNC niche, that magical overlap of what you're good at, what people want, and what the market doesn't already have in a hundred variations.

Why Niche Matters (Especially in CNC)

Imagine you're a customer looking for a custom sign. You go online and find ten listings. One says, "Custom CNC Wood Signs." Another says, "Custom Farmhouse Signs with Vintage Fonts & Routed Borders (Made in Maple or Walnut)."

Which one gets the click?

The niche one. Every time. *I'm assuming that's exactly what the customer was looking for.*

Why? Because specificity builds trust. It signals expertise and makes the buyer feel like you understand their exact need or style. In CNC work, where tools are common and materials are shared, your niche is what sets you apart from the generalistis out there.

Niching helps you build a recognizable style, charge more for specialized work, create repeatable processes, attract the right kind of customers, and confidently say no to the wrong ones. Get especially good to saying ***NO***.

You might be wondering what *CNCROi.com*'s niche really is. Our focus is on industrial services where there is a clear and measurable Return On Investment. Ironically, because of NDAs, what you see on my website often looks quite the opposite. I know, it is a classic case of 'do as I say, not as I do.' The truth is, running a business is far more complex than writing a book about it.

The Myth of "Generalist = Versatile"

It's tempting to think that offering more will bring in more business, CNC routing? Sure. Laser cutting? Absolutely. 3D printed cosplay gear, wedding signage, metal parts, corporate awards, dog bowl stands, light sabers? Why not.

But that's not a business. That's a panic attack waiting to happen. And the stress only builds-on from there.

Being a generalist feels safe, but it often leads to inconsistent quality, constant context switching, pricing confusion, and a brand identity that boils down to "I do stuff." Having a niche doesn't mean you can't branch out, it means your core offer is tight. Clean. Understandable. Marketable.

Step 1: Audit Your Skills and Machines

Start with a clear, honest look at what you do well, and what your machines do best. Ask yourself:

What materials do you enjoy working with? Which machine do you know inside and out? What types of projects make you lose track of time (in a good way)?

Your niche should build on your strengths. If you hate sanding, steer clear of hand-finished walnut plaques. But if you love mechanical puzzles, maybe you're the next big name in laser-cut escape room props.

Step 2: Research What the Market Wants

You don't need to invent a new category, you just need to find the intersection of demand and differentiation.

Start by researching. Browse Etsy, Instagram, Pinterest and even my own website, *CNCROi.com*. Dive into niche forums like cosplay, overlanding, or home renovation. Use Etsy and Google search suggestions by typing phrases like "custom CNC ____" to see what people are actually searching for. Look at what's already popular, then ask yourself: how could I do this better, faster, or in a more unique material?

Keep an eye out for underserved niches, like marine-grade cabinetry, educational math manipulatives, or parts for vintage cars. Again, anything automotive is something I tend to avoid but maybe that's where you shine.

Focus on high-margin products that benefit from customization, and target repeat customers such as event planners, interior designers, or other fabricators.

Marketing to poor people or trying to get low margin work is a great way to be busy without making any money to support your passion. Never compete on price, it's a loss-loss proposition filled with empty and broken promises.

Step 3: Find Your Overlap Zone

Now combine three things: what you enjoy, what you can produce reliably, and what people will pay for. That's your sweet spot. Your job now is to optimize in that direction.

Let's say you love lasers, you're skilled with Illustrator, and you have a thing for mid-century modern design. Great! Start a line of retro-styled laser-cut clocks. Or maybe you're a welding nerd with a plasma table and a passion for off-road gear, perfect. Launch custom roof racks for 4Runners.

Avoid niches that demand entirely new skill sets unless you're genuinely committed to learning them. Your niche should energize you, not bury you under YouTube tutorials and burnout. Believe me, been there, done that.

Step 4: Name It, Frame It, Market It

This is where niche becomes brand. Start putting language to what you do:

Instead of: "I make stuff on a CNC router."
Say: "I make custom wall panels for modern interiors using sustainably sourced wood and clean geometric patterns."

Or:

Instead of: "I do 3D printing."
Say: "I design and print hyper-detailed fantasy miniatures for tabletop gamers who care about fine detail."

When you describe what you do with specificity and style, people remember you. They refer you. They get what you offer. And your niche starts to sell itself.

What If the Niche Feels Too Small?

Good. That means it's focused.

Here's a secret: a tight niche doesn't limit your business, it grows it faster. When you become the go-to shop for one specific type of work, you build credibility, attract followers, collect testimonials, and, most importantly, gain momentum.

Later, you can expand into adjacent products like matching accessories, branch into new markets like moving from Etsy to commercial clients, or explore limited editions and collaborations.

But start focused. You can always zoom out. It's much harder to go the other way.

Thanks to the internet, even the smallest niche can turn into millions of dollars and thousands of customers worldwide. A custom CNC business that once only served its local community can now reach a global audience. Never underestimate how large your customer base can be beyond your local market. Shipping has never been easier.

What If You Like Doing Lots of Things?

Fair. That's a maker trait.

Try this strategy: have one niche business, and a playbench for everything else.

Keep your product line or service offering focused, but still allow yourself time to prototype, tinker, and chase creative whims. You never know when one of those side projects becomes your next niche.

But don't let your creative wanderlust hijack your marketing.

Branding the Niche: Visual and Verbal Style

Your niche is more than what you make, it's how you present it.

Use consistent fonts and color palettes, take great photos in good light, and give your products descriptive names. Tell the story of your process, and share behind-the-scenes looks at your tools, setups, and iterations.

People don't just buy your product, they buy your vibe. And your vibe should align with your niche.

Pricing Within a Niche

Niches allow for premium pricing because you're specialized, your work feels custom or high-skill, and your clients feel seen.

Just make sure your pricing reflects everything involved: material and tool costs, setup and design time, finishing or shipping, and most importantly, your growing expertise.

Don't race to the bottom. That's where generic makers go to die. Also, be careful of who or how you tell others if you happen to find a highly profitable niche. Remember, there are vultures everywhere without any creativity just looking to undercut your niche for their own riches.

Final Thoughts: Your Niche Is a Gift

Your CNC niche isn't a limitation, it's a megaphone. It allows you to speak clearly in a noisy market. Especially with YouTube, your audience in a niche can be globally relatively quickly at no real cost to you.

When you find your niche, marketing becomes easier, your work gains focus, your pricing improves, you attract better-fit customers, and you start getting known for something. And once you're known for something, you can build anything.

So the next time someone asks what you make, don't just say, "I do CNC." Say, "I make [your niche here], and my machine just happens to be digital."

CNC is a means to an end, don't let it define what you actually do or the services you provide. And, make sure you don't define yourself as a CNC ____ business, you are leaving a lot of money on the table by going this route.

Laser cut polycarbonate keyboard template, then sand blasted.

Appendix E: Marketing Your CNC Business Online: Lessons from CNCROi.com

If your CNC machine is running but your inbox is quiet, you don't have a shop problem, you have a marketing problem.

In the digital fabrication world, where thousands of makers compete for attention and dollars, having an online presence isn't a luxury. It's survival.

Enter *CNCROi.com*, a website that should be required reading for anyone serious about growing a fabrication business online. It's not flashy. It's not trendy. But it works.

CNCROi.com has quietly become one of the most effective content-driven CNC business platforms on the web, with over a thousand blog posts, thousands of videos, and a steady stream of qualified inquiries.

Why does *CNCROi.com* succeed? Because it knows how to leverage SEO without compromising authenticity, and it applies specific tactics that any CNC business can use to turn clicks into contracts.

The CNCROi.com Strategy: Authority Through Content

The backbone of *CNCROi.com* is content: a relentless, structured, and strategic stream of blog posts, videos, and tutorials.

There are no gimmicks, no desperate sales pitches, just project after project, well-documented and clearly explained. At least to me.

This content machine does three things exceptionally well: it educates the customer, builds trust with prospects, and dominates search results in a niche few truly understand. How is AI going to affect this? Haven't a clue, but AI isn't creative yet so I think I'm safe.

CNCROi.com isn't chasing virality. It's playing the long game, over a decades, focusing on organic visibility, lasting credibility, and steady lead generation over time.

Why Blogging Still Matters (Especially for CNC)

In a social media world, blogging might seem old-school—but it's a powerhouse for SEO.

Every blog post on *CNCROi.com* creates a new entry point for search engines and prospective clients. Someone Googles "laser engraved stainless steel signage," and they land on a detailed post with video, context, and a clear call to action.

That's the game I play. Not likes, not shares, but searchable, evergreen content. Each post uses long-tail keywords naturally, includes descriptive titles and URLs, embeds relevant video and photos, and solves a real-world problem.

This is how you rank. This is how you build traffic that converts.

There is absolutely no question in the customer's mind that I can do the work, I get paid upfront and deliver right to their doorstep or loading dock. Easy and seamless.

Video: Not Just for YouTube Celebrities

CNCROi.com integrates video into nearly every project showcase, not for entertainment, but for documentation.

Videos offer proof: that real work is being done, that the process is understood, and that the machines are fully capable. This social proof is vital in my opinion.

You don't need studio lighting. You just need a smartphone, a tripod, and a steady voice. Talk through your setups, show your mistakes, and walk viewers through the job from file to finish. As a bonus, video keeps visitors on your site longer, boosting your SEO in the process.

I know there's room for improvement. I could comb my hair, iron my shirt, or even shave before starting a project video. But the truth is, customers don't care about that. They have a need, I have the skills, and that is what matters, no matter how rough my approach may look.

Site Structure and Flow: Built for Browsing

CNCROi.com's site structure is deceptively simple. It organizes content into clear categories such as materials, processes, and industries, with internal links connecting related jobs. Navigation is straightforward with a sidebar and archives, and the hierarchy makes sense: services lead to the portfolio, which leads to the contact page.

There is also a contact option on every page, making it easy to get in touch.

This structure keeps users engaged. They click deeper, stay longer, and absorb more, which sends Google positive signals through better time-on-site and lower bounce rates.

Your site should follow the same principles. Make it mobile-friendly, fast-loading, and easy to navigate with a clear menu.

Display your contact information prominently and include calls to action that guide users without annoying them.

You don't need professional photography.

Keywords Without Keyword Stuffing

CNCROi.com doesn't shout "CNC! Laser! Engraving!" in every sentence, it writes like a human. Keywords are there, but they're natural and never distracting. The key is writing about real projects using the language your customers actually search for.

Think in terms of titles like "Custom aluminum tags for military application," "Laser engraved birch plywood for architectural model," or "High-precision industrial control panel engraving."

Write blog posts like you're answering a customer question, because that's exactly what Google rewards.

Backlinks and Authority: Earned, Not Bought

With enough high-quality content, people naturally link to *CNCROi.com*, from forums discussing laser engraving to educational institutions citing case studies, and customers sharing their completed projects. These backlinks signal trust to Google, and trust equals rank.

You can earn backlinks too. Write guest posts for industry blogs, share your project results with vendors who might repost them, and submit your work to fabrication showcases.

The Blog as Sales Funnel (Without the Funnel Hype)

CNCROi.com doesn't push hard sells, there's no countdown timer, no pop-up promising a free ebook in exchange for your email. The sale happens subtly.

A visitor lands via search, browses multiple blog posts, sees video proof and testimonials, and eventually ends up on the contact page.

That's a funnel, just not the annoying kind. Each piece of content gently builds trust. By the time they reach out, they already believe *CNCROi.com* can handle the job.

Content Consistency > Social Virality

CNCROi.com doesn't chase temporary trends: it just posts regularly, week after week, year after year, decade after decade.

That kind of consistency does more for long-term lead generation than any viral post ever could.

You don't need to go big; you just need to show up.

One blog post a week means 52 new search-friendly pages a year. One video a month gives customers 12 new reasons to trust you.

SEO rewards consistency, and so do customers.

What You Can Do Today?

Start a blog on your website and document every project with photos and a few paragraphs. Share what you did, the tools used, and any lessons learned. Record short videos of your workflow and organize content by material, industry, and tool so visitors can find relevant posts easily.

Do not worry about appearance or background at first; customers care more about results than production quality, which you can improve over time.

Use natural keywords based on customer questions and include internal links between posts to guide visitors and help Google understand your site.

Make it easy for visitors to contact you and add calls to action like requesting a quote or scheduling a consultation. Stay consistent for at least six months to see meaningful results.

Over time, this approach can turn casual visitors into loyal customers and position your business as an authority in your niche." Oh yeah, get your own website domain and watermark all your content with it so that content thieves are working with you, not against you online.

Final Thoughts: Marketing Like a Maker

The best marketing does not feel like marketing. It feels like teaching. It shows what is possible. It delivers value first and lets real expertise speak for itself. Visibility is not luck. It is intentional architecture.

That said, *CNCROi.com* hit a major roadblock I could not have predicted when I first wrote this book. On August 29, 2025, the U.S. de minimis exemption was cancelled. At first glance, that might sound minor. It was not.

As of this edit on January 1st, 2026, *CNCROi.com* has not shipped a single package to the United States. Overnight, I lost my entire American customer base. That market represented most of my business. I am based in Canada and customs quickly became unworkable for a custom shop.

Everything I wrote above is still true. The strategy still works. The problem now is a mix of logistics and politics, two areas I have no control over. I now spend time declining paid work because cross-border shipping has become an absolute mess. Imagine customers offering you money and having to turn it down. Over and over... and over again.

The most logical move is to grow my Canadian customer base, which I am slowly doing. But the reality is that I spent ten years building a business focused almost entirely on the U.S. market. COVID also wiped out craft shows, which used to provide strong holiday and year-round revenue. That income stream never came back afterwards.

This is not meant to be a downer, but business risk does not always arrive as a small setback. Sometimes you are not hit by a car. You get run over by a freight train you never saw coming, nor your customers who depended on you for years to produce exactly what they needed.

My solution, while I slowly rebuild during a global depression, was to take a job as a welder. The goal is to learn new skills I can eventually bring back into my business. Switching gears is not fun, but in many ways it feels like the third year of welding school I wish I had taken. See Appendix G, page 346.

Appendix F: Metrology & Inspection

Precision fabrication demands more than just cutting, printing, or welding parts. It requires verifying that what was made actually meets the design intent.

Metrology is the science of measurement, and in the CNC world, it's your best friend for:

- Ensuring parts fit and function
- Validating repeatability and accuracy
- Reducing waste and rework

Each digital fabrication method introduces unique variables: taper, warping, shrinkage, distortion, etc.

CNC Plasma Cutting

Key Inspection Challenges

- Tapered edges from arc angle or height variation
- Dross buildup on underside
- Heat warping or discoloration
- Hole deformation

Keep in mind that these challenges vary depending on the cutting setup and can differ significantly when using a downdraft table versus a water table.

Measurement	Purpose & Tools
Kerf Width	*Fit accuracy, tab/slot designs. Feeler gauges, calipers*
Edge Bevel/ Taper Angle	*Quality control, post-process fitment. Bevel gauge, square, protractor*
Dross Thickness	*Clean-up time, safety. Visual, caliper, pick tool*
Hole Roundness	*Fastener fitment, distortion check. Pin gauges, digital calipers*
Surface Oxidation	*Weld or paint readiness. Wire brush test, visual*

CNC Waterjet Cutting

Waterjet cutting uses a high-pressure stream of water, often combined with abrasive particles, to cut a wide range of materials without generating heat.

Because it is a true cold-cutting process, it is well suited for heat-sensitive or distortion-prone materials such as composites, aluminum, and stainless steel. When abrasives are not used, the process is also effective for cutting soft materials like foam.

While waterjet cutting delivers excellent edge quality and exceptional material versatility, it also introduces a distinct set of inspection challenges. Addressing these issues is essential to maintain dimensional accuracy and reduce the need for secondary finishing.

Key Inspection Challenges

- Stream lag taper in thick materials
- Uneven pierce zones or delamination
- Striation roughness on lower cuts
- Warping from unsupported thin sheet

Keep in mind that these challenges can vary significantly depending on the power of the waterjet stream.

Measurement	Purpose & Tools
Taper Angle	*Fit-up in assemblies. Calipers (top/bottom), taper gauge*
Surface Finish Quality	*Post-processing and spec compliance. Profilometer, comparator blocks*
Pierce Zone Defects	*Cracking in brittle materials. USB scope, dye penetrant*
Hole Roundness	*Fastener or insert compatibility. Bore/pin gauges*
Sheet Flatness	*Material handling, downstream issues. Dial indicator, straightedge*

CNC Laser Cutting and Engraving

Laser cutting and engraving use a focused beam of light to melt, burn, or vaporize material with high precision, making it a go-to process for both decorative and functional parts. This method excels at producing fine details, sharp edges, and intricate markings across a wide range of materials. From acrylic and wood to metals, laminates, and even some fabrics.

Its non-contact nature reduces mechanical stress on the workpiece, allowing for delicate features and minimal fixturing. However, the thermal nature of laser processing introduces specific inspection concerns, especially when balancing speed, edge quality, and surface finish.

Excess heat, poor focus, or material variation can quickly lead to quality issues.

Key Inspection Challenges

- Edge discoloration or charring
- Inconsistent cut depth or taper
- Engraving depth variation
- Material scorching or back reflection

Measurement	**Purpose & Tools**
Taper Angle	*Fit-up in assemblies. Calipers (top/bottom), taper gauge*
Surface Finish Quality	*Post-processing and spec compliance. Profilometer, comparator blocks*
Pierce Zone Defects	*Cracking in brittle materials. USB scope, dye penetrant*
Hole Roundness	*Fastener or insert compatibility. Bore/pin gauges*
Sheet Flatness	*Material handling, downstream issues. Dial indicator, straightedge*

CNC Router Machining

CNC router machining is a subtractive process that uses high-speed rotary tools to cut and shape materials like wood, plastics, composites, and soft metals with precision and repeatability. Known for its versatility, broad material compatibility, and large work area, the router excels at contouring, pocketing, and profiling in sheet and panel goods, often in a single setup.

However, maintaining accuracy across wide surfaces and managing tool dynamics can present unique inspection challenges, particularly when dealing with natural material inconsistencies, tool wear, or complex multi-sided operations that demand precise registration and alignment.

Key Inspection Challenges

- Z-depth variation across spoilboard
- Tool deflection on long reach cuts
- Tear-out or poor edge finish
- Double-sided registration errors

Keep in mind that bit selection plays a critical role in the final result. A roughing bit will produce a noticeably different edge profile and surface quality than a finishing bit, even when run at the same feeds and speeds.

Measurement	**Purpose & Tools**
Cut Depth Accuracy	*Ensures through cuts, pocket integrity. Depth gauge, caliper*
Spoilboard Flatness	*Surface consistency. Dial indicator, flycut result*
Bit Deflection Taper	*Diagnoses chatter or poor setup. Caliper (top vs. bottom edge)*
Edge Finish Quality	*Clean parts, minimal sanding. Fingernail test, visual*
Double-Sided Alignment	*Symmetry on flipped operations. Dowel pins, centerline caliper*

Welding

Welding joins materials, typically metals or thermoplastics, by melting the base material and often adding a filler to create a strong, permanent bond. Whether done manually or via automated systems, welding is critical in structural fabrication, and mechanical joining across a wide range of industries.

Because it involves high heat, material flow, and solidification, weld quality can vary widely depending on technique, setup, joint design, and material properties. Inspection plays a vital role in ensuring structural integrity, cosmetic standards, and code compliance, often determining whether a weld is simply acceptable or requires costly rework.

Key Inspection Challenges

- Undercut or overfilled welds
- Porosity or cracking
- Heat distortion
- Incomplete fusion or penetration

Keep in mind that filler material selection, amperage, and travel speed all have a significant impact on the quality of the finished weld.

Measurement	**Purpose & Tools**
Fillet Size /Profile	*Strength, compliance with WPS. Fillet gauges, weld gauge*
Undercut /Overlap	*Visual and structural quality. Weld gauge, dye penetrant*
Penetration Depth	*Fusion assessment. Macro etch sample, UT, radiograph*
Post-Weld Distortion	*Maintains tolerances. Angle finder, straightedge, calipers*
Weld Surface Defects	*Cosmetic or fatigue risk. PT, MT, USB scope*

3D Printing (FDM/SLA/SLS)

3D printing, also known as additive manufacturing, builds parts layer by layer, offering flexibility for prototyping and low-volume production. FDM is cost-effective and accessible, SLA provides high-resolution detail, and SLS produces durable, support-free prints.

While these methods reduce the need for fixturing and tooling, they introduce unique inspection challenges related to material behavior and machine calibration.

Key Inspection Challenges

- Dimensional drift in XY and Z
- Warping or bed adhesion failure
- Layer delamination
- Stringing, over- or under-extrusion

Keep in mind that filaments advertised as the same material are not necessarily identical across manufacturers. These differences, along with build plate heating and overall movement speeds, can significantly affect the results you achieve.

Measurement	**Purpose & Tools**
Dimensional Accuracy (XY)	*Assembly fit, press-fit components. Calipers, test cubes*
Layer Consistency	*Surface finish and strength. USB microscope, macro photos*
Warping	*Flatness of base, bed adhesion. Ruler test, dial indicator across bed*
Infill Integrity	*Structural strength. Visual cutaway, destructive test*
Surface Artifacts	*Cosmetic and functional defects. Visual inspection, test swatches*
Shrinkage Post-Cure (SLA)	Dimensional scaling. Caliper before/after cure

Your Metrology Toolbox

Accurate inspection starts with using the right tools for the job. Each manufacturing process introduces its own sources of variation, and no single instrument can capture every critical feature. The tools below represent a practical inspection kit used across machining, cutting, welding, laser processing, and additive manufacturing to verify dimensions, surface quality, alignment, and process consistency before parts move downstream.

Digital Calipers (0.01 mm / 0.0005"):
General part dimensions, hole spacing

Dial Indicator + Magnetic Base:
Spoilboard flatness, Z-axis tram

USB Microscope:
Layer inspection, engraving contrast, weld surface defects

Weld Gauge Set (fillet, undercut):
Weld size conformity and visual flaw checks

Profilometer / Comparator Blocks:
Surface roughness on jet/laser edges

Bevel Gauge:
Plasma, laser, or waterjet edge angle

Go/No-Go Gauges:
Hole fit checks across all processes

Depth Gauge:
Pocket depth, engraving depth, or layer thickness

Dye Penetrant (PT):
Surface crack and porosity detection in welding or waterjet

Stair-Step Focus Block (Laser):
Dialing in engraving and cut focus

Shrink Test Prints (3D):
Scaling correction for resin-based 3D printers

Best Practices

Start measuring early in the workflow, inspection should begin before the first cut is made. Early checks help catch issues before they lead to rework or waste.

Track your results consistently using job logs or digital quality control sheets to maintain accuracy and accountability. Use process-specific inspection checklists to ensure all critical steps are covered.

Clean and calibrate your measuring tools regularly to ensure reliable readings. Promote consistency by using the same tool, method, and operator whenever possible, this reduces variability and improves data reliability.

Remember, tolerances vary by process.

What's acceptable in woodworking may not be suitable in metalworking, so always align expectations with the material and application.

A bold, massive live edge statement sign (reference tape still on).

Appendix G: So you want to be a Welder?

Due to the de minimis rules being cancelled for shipments from Canada (where I am based) to the USA (where most of my customers *were*), business slowed down dramatically. I decided to use that slowdown as a chance to get my first traditional job at the age of forty-nine.

Unexpected opportunities tend to show up at the worst possible time. That is actually how I ended up in college in the first place. After COVID and the rise in interest rates, my business took another hit, so I returned to school to learn a skilled trade that has always been on my bucket list.

I always grew up thinking the worst thing that could happen in life was getting a real job, so waiting this long to try it felt like the right move. An anti-bucket list if you will.

After earning my two-year welding diploma, I applied for a handful of welding and fitter roles, and one company hired me to do exactly that along with quite a bit more. I kept my resume simple because of my unconventional background. Most companies are not looking for a welder who has spent years abroad teaching ESL, running businesses, working as an entrepreneur and a borderline senior citizen.

At school, our instructors reminded us that a welding career is something you move quickly through. You get in, earn your money, and get out before the physical wear and tear catches up. It's a brutal but fun trade!

My original plan after graduation was to offer welding services independently. I even had a few projects lined up, but the economy shifted again and those opportunities went up in smoke.

Student VS Real Welding

I am fortunate to have an amazing boss, an incredible shop boss, and a group of colleagues who are genuinely great to work with. They have been endlessly patient while I learned how welding actually works in the real world.

Niagara College took me from basically zero experience to about twenty-five percent after two years, but the welding conditions there were mostly ideal.

Machines were always set up properly, materials were consistent and predictable, the plans were standardized, and everything was controlled. None of that matches the reality of a working shop.

It did not take long for me to learn that the simple 1, 2, 3, and 4 welding positions become 1.4, 2.1, 3.6, and "good luck with that one" once you are on the job. Flat is suddenly a relative term. Wire behaves differently. Settings that worked every time in school are nowhere near correct in production. It was a REAL challenge at first.

Our instructors used to say that ninety percent of welding is not actually welding. I never really understood that until now, but it is absolutely true.

If a part needs to be galvanized after you build the assembly, you start by sealing every threaded hole with silicone. I did not know that. I also did not know that this can take hours. When the parts come back from the galvanizing shop, you remove all that silicone and lightly chase every thread to clean out whatever got in there.

If something needs paint, that is another adventure. You had better be ready to spend two days with a wire wheel on an angle grinder removing industrial paint from a large welded assembly. Then you clean everything with brake cleaner, a product welders dread, before the new coating can go on after doing additional welding on it.

All of this opened my eyes. Welding is not as simple as prepping a joint and striking an arc. There is rust removal, paint removal, cleaning, scraping, caulking, galvanizing work, fixing threads, and a long list of other jobs that make it possible to actually weld in the first place.

Learning to work with imperfect material

Unlike school, real world material can be warped, bent, pitted, or cut wrong by you or someone else. Our teachers used to tell us that being within one eighth of an inch, roughly three millimeters, was acceptable.

I used to be amazed at how loose that sounded, but now I understand. In the real world, that is often the goal, and sometimes the standard needs to be even tighter.

Surface preparation becomes a completely different challenge too. In school, everything was clean and consistent. On a job site, you might be working with parts that were sandblasted months ago, coated in some unknown fluid, or covered in a year's worth of rust because they sat outside until the customer finally decided they needed the work done.

The biggest wake up call for me was dealing with imperfect fits. You might need to weld two pieces together only to discover a quarter inch gap staring back at you. At that point you figure it out. You take a piece of scrap metal that is close to the right thickness and use it as a makeshift backer. You adjust, you adapt, and you make it work. That is real world welding in a nutshell.

How fabrication teaches problem solving

I may work in a welding shop, but I still do not call myself a welder, even though I now hold CWB tickets. I probably know five percent of what an experienced welder knows. Their skill level is unreal, especially when you see them repeat the same high quality work every time.

I am learning a huge amount from them, but you simply cannot compare my three years of experience to someone who has been welding for thirty. What I can say is that it is a privilege to have access to their knowledge, and yes, it truly is an honor to be in their presence.

One time I was struggling with a section of railing. A veteran welder walked by, watched me for a minute, and asked why I was not just building a simple jig and welding it to the table. In hindsight it was obvious, but to a newcomer like me it was a complete lightbulb moment.

So I grabbed some angle iron, welded a vertical and horizontal setup right onto my table, and suddenly the job became twice as easy. Grinding went smoother, welding the tube rails was far more controlled, and the results were noticeably better. I could lean on the work instead of trying to hover over it.

After about a month on the job, I asked my shop boss why so many people think welding is easy and look down on it. I found it to be anything but easy. It is physically exhausting, mentally demanding, and there is real problem solving involved every day. In a metal fabrication shop, nobody coasts until lunchtime like you can do in an office.

I have gone to university, and I can say without hesitation that a desk job pays more and is far less taxing. After working in fabrication, it is clear to me that the real thinking often happens in the trades.

Why skilled trades are looked down upon makes no sense, especially now that I have experienced both worlds.

Physical demands and shop rhythm

Going back to that railing job, my production time was around three hours. After I built the jig, I cut that time almost in half and felt pretty good about it. Then my shop boss told me the target was thirty minutes. I could not understand how that was even possible.

To figure it out, I wrote down the start and stop times for every step of the process right on my welding table. That is when I realized the actual welding and surface prep took me about fifteen minutes. That part was fine. The problem was that the rest of my time was spent fixing my own welds. Warping, grinding, redoing beads, correcting porosity, and cleaning things up after the fact was eating up hours. Pipe based railing isn't easy welding!

A skilled welder barely has any fixing to do. They take the time to set up properly, dial in their position, and focus on technique. They are not rushing the bead. They know that the first weld is the most important weld. Meanwhile, I was chasing the clock and creating twice as much work for myself in the process.

I also learned very quickly how physically demanding a full day of welding can be. Your body ends up twisted into awkward angles, and there is always something blocking your perfect position. You are on your feet from start to finish, dealing with heat, burns, and sudden cold whenever you step away from the torch. Grinding alone can wear you down fast. Welding is basically an intense workout done using all the worst ergonomics possible.

The biggest lesson for me was learning to pace myself, respect the physical demands of the job, and understand that speed comes from precision, not rushing.

Veteran welders know this instinctively, which is why they always look so calm and almost slow in how they move. Now I understand it. By going slow, they are actually going fast. It makes perfect sense once you see the difference.

Safety awareness and “real-world” lessons

Welding is hot in ways you do not fully appreciate until you start doing it every day. I once had my glove catch on fire while welding an assembly and did not notice until another welder walked over because he smelled burning leather.

Then there is FCAW slag. It looks solid and harmless, but somehow it feels hotter than molten metal itself. When you chip it, it always finds the one spot you forgot to cover. I made the mistake of opening my welding jacket a little around the neck because I was overheating. The slag landed there and slowly worked its way down. That one took a while to heal and left a good scar.

PPE matters, but sometimes even that is not perfect protection. I have had flying sparks travel under my hood, squeeze between my safety glasses and my face, and melt into the safety glass lens instead of my eye. I was lucky.

Welding is hazardous for your lungs, eyes, skin, fingers, and anything else that ends up in the line of fire. Little oversights lead to burns, cuts, smashed fingers, and a collection of small injuries that remind you to stay alert. Welding is a mental sport. You have to be fully aware all the time or a small mistake becomes a big problem.

If you want a visual reminder of what metal fabrication does to a body, talk to a veteran welder. They will show you the scars to prove it. When I showed a welder the burn on my neck from slag, he showed me one on the back of his neck. How slag got there is anyone’s guess. He also told me that sometimes slag lands on your lips, and your instinct is to cool it with your tongue. That usually ends with a burned lip and a burned tongue. Welding life has its own special kind of fun.

Another constant safety concern is material flow in the shop. Parts come in, assemblies go out, and all of it is moved with forklifts and overhead cranes that operate at surprising speed. You have to stay aware of your surroundings at all times. These machines do not forgive mistakes, and being in the wrong place at the wrong moment can end very badly.

Real world welding has taught me that safety is not a rule book. It is a habit, a mindset, and something you must respect if you want to stay in the trade. Hang-out with veterans, they'll see what you don't and help you avoid it.

The shock of production timelines

Experienced welders move with precision and purpose. Their motions are deliberate and efficient, with almost no wasted energy when they position parts, prepare joints, or set up a weld. Everything they do has a clear reason.

One welder at our shop often has to weld in front of supervisors and clients. For me, that would be a high pressure moment because every detail of the weld is inspected immediately afterward. For him, it is just another day. He has dealt with tight deadlines and demanding customers for so long that the pressure does not faze him anymore. He understands speed, quality, and compromise instinctively where I still think too much.

Respect for the trades

Welders and fabricators rarely get the credit they deserve for the complexity of their work. Metal fabrication is demanding in every possible way. It requires skill, awareness, problem solving, physical endurance, and the ability to think ahead while working in tough conditions. The people in this trade are some of the hardest working and smartest individuals I have ever met.

When I was in university, one of my professors held several PhDs and taught business, yet he had never run a business in his life. He was also the worst instructor I ever had. Despite this, he earned far more than any welder, even though he would not last a single day doing the work I do now. He was nowhere near as capable or efficient as the tradespeople I work with every day, yet he would be treated with far more respect in society than the welder assembling I-beams to build a bridge outdoors in the middle of winter.

Growing up, the message was always the same. Kids who were not smart went into the trades, and the smart ones went to university for clean, high paying careers. Now that I have lived both sides of that belief, I can say with confidence that the attitude should be reversed. The trades are not a fallback. They are a discipline that demands intelligence, awareness, and years of experience.

Jeremy is close to my age and has a child 2 years older than mine, but he chose a different path. He left school early for metal fabrication, learned everything hands on, and built his career step by step the hard way. Thirty years later, we work in the same building, yet he is my boss, with more skill, higher pay, and far more respect than I could claim. ***That is exactly how it should be. He more than deserves it.***

Other Books by Jonathan Cantin

Mastering CNC and Digital Fabrication
The Ultimate Practical Guide to CNC Routing, Laser, Plasma, Waterjet, 3D Printing, and Welding

366 pages / 2025
ISBN 978-1-896369-53-2

This book is the foundation from which the book you are reading right now came from. Together, they form a one-two punch of CNC knowledge I wish I had 20 years ago.

Mastering CNC and Digital Fabrication is a no-nonsense guide to real-world making. Whether you're prototyping, producing, or just trying not to start a fire, it’s about working smarter, not just harder.

You’ll get hard-earned lessons from decades in CNC routing, laser cutting, 3D printing, welding, and more. Learn to design for real-world tolerances, choose the right materials, and navigate machine types, from CO_2 lasers to resin printers. Skip common mistakes, optimize your workflow, and make the most of automation and AI.

This book cuts through the fluff with case studies, practical tips, and a glossary that actually makes sense if you can think outside the kerf.

It’s for soon-to-be CNC operators, makers, shop owners, and designers who live on the shop floor, not in theory.

Because making it work always beats hoping it does.

CNCROi.com V5: Think it, Build it, Sell it!

504 pages / 2015

ISBN 978-1896369525

Two years in the making, this teaches you how to design for CNC machines and successfully market your creations.

This book offers practical advice for designers and business owners, written in an easy-to-understand, non-technical style. It's a great resource for those looking to learn about CNC machines and improve their business.

CNCKing.com Volume 4: Rise of the CNC - Ultimate CNC Design Course

456 pages / 2013

ISBN 978-1896369518

This book covers designing for CNC Table Routers, 3D Printers, and Laser Cutters. Over 450 pages of projects, expert interviews, and tips to help you get the most from your CNC machines.

HowToChinese.com: A Practical Mandarin Chinese Course

412 pages / 2012
ISBN 978-1896369501

Learning a new language is fun with a book that guides you every step, written from a student's perspective!

Xiaobei's passion for teaching Mandarin comes to life through her video series, perfect for beginners and advanced learners alike.

Key Features:

Step-by-step process with building chapters

Exercises, real-life dialogues, and mini-dictionaries

Pinyin, Simplified Chinese, and cultural insights

170+ photos from China

Comprehensive video series to boost fluency

Focused on Putonghua, this book offers a complete learning experience, with bonus materials and multimedia support. Written by Xiaobei and Jon, it's the perfect start to mastering Mandarin.

WoodMarvels.com: V3 - Evolution of Wooden Designs
200 pages / 2011
ISBN 978-1896369495

This book features woodworking projects across five skill levels with 3D step-by-step assembly instructions. Our unique philosophy: all blueprints are measurement-free, allowing you to scale projects based on wood thickness. Included models from WoodMarvels.com:

Note: Does not include DXF or EPS files.

WoodMarvels.com, Volume 2: Laser Cutting Plans
200 pages / 2009
ISBN 978-1896369464

This WoodMarvels.com book includes blueprints and 3D assembly instructions for models across four levels:

Level 1: Simple creatures and everyday objects
Level 2: Vehicles, functional items, and themed designs
Level 3: Advanced projects with unique concepts
Level 4: Large-scale models and mechanical designs

2847 A.D.: Solar Horizons
214 pages / 2009
ISBN 978-1-896369-47-1

In 2847, humanity struggles to survive after near-extinction. Extraterrestrial colonies thrive, and trade is essential.

A Professor and Excur, the most advanced computer, uncover an unexploded neutron sphere in Zulu territory, risking Earth's destruction. Joined by three mining Alabarsi, they seek answers. What's Project Baptistina, and how is Glass Lake tied to humanity's possible extinction?

Transforming your life from Blah to Legendary!
140 pages / 2009
ISBN 978-1-896369-45-7

The biggest obstacle to the life you want is yourself.

Creating lasting change requirestime, new habits, and a shift in attitude. True success begins with understanding and strengthening yourself.

If you're serious about living a fulfilling life, this book is a great first step. You only have one life, make it count!

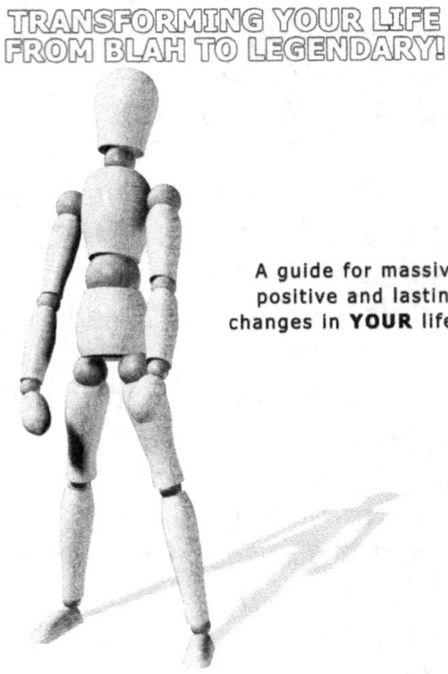

www.ingramcontent.com/pod-product-compliance
Lightning Source LLC
LaVergne TN
LVHW052226200926
848481LV00001B/227

* 9 7 8 1 8 9 6 3 6 9 5 4 9 *